'A glittering cloud of Inupiaq, Icelandic, compound Māori, Finnish, Scots, Thai, Hebrew, American Sign Language – this book is a miraculous snow bank of niveous names and knowledge as delicate and multifaceted as the flakes it celebrates.'
– Dan Richards, author of *Climbing Days*

Nancy Campbell is a poet and non-fiction writer described as 'deft, dangerous and dazzling' by the former Poet Laureate, Carol Ann Duffy. Her work has engaged with the polar environment since a winter spent as Artist in Residence at the most northern museum in the world on Upernavik in Greenland in 2010. Her books include *The Library of Ice: Readings in a Cold Climate*, *Disko Bay* and *How to Say 'I Love You' in Greenlandic*. She was appointed Canal Laureate by The Poetry Society in 2018, writing poems for installation across the UK waterways from London Docklands to the River Severn, and received the Ness Award from the Royal Geographical Society in 2020.

FIFTY

WORDS

FOR

SNOW

FIFTY WORDS FOR SNOW

NANCY CAMPBELL

Elliott&Thompson

First published 2020 by
Elliott and Thompson Limited
2 John Street
London WC1N 2ES
www.eandtbooks.com

ISBN: 978-1-78396-498-7

Photographs by Wilson Bentley (1865–1931), the first known photographer of snowflakes.

Permissions:
Page ix: Traci Brimhall, excerpt from 'Dear Eros' from *Come the Slumberless to the Land
of Nod*. Copyright © 2020 by Traci Brimhall. Reprinted with the permission of The
Permissions Company, LLC on behalf of Copper Canyon Press, www.coppercanyonpress.org.

Page xv: 'Digging' from *Death of a Naturalist*, © Seamus Heaney (London: Faber and Faber,
1966), reprinted by permission of Faber and Faber Ltd.

Page 111: ASL image, William G. Vicars, www.lifeprint.com

9 8 7 6 5 4 3 2 1

A catalogue record for this book is available from the British Library.

Cover design by Jo Walker
Typesetting by Marie Doherty
Printed in the UK by TJ Books Limited

for Anna
who lost all her words
then began to find them again

Even snow
knows it's unclean. Each flake makes its own geometry
around dust, where everything begins.

— Traci Brimhall, 'Dear Eros'

Contents

Contents

Contents

Prologue

A few winters ago I rented a former Salvation Army meeting house in the north of Iceland for a few months. Since the snows of Siglufjörður don't usually melt until April, I soon learned how the inhabitants of this small fishing town distinguish themselves from their neighbours in Ólafsfjörður, another small fishing town beyond the mountains. Folk in Ólafsfjörður do not clear the snow from the paths leading to their homes. In Siglufjörður the sweeping of snow is a social duty. I lived alone, and it was some distance from my front door to the road; the snow lay in waist-high drifts, and kept falling. I'd gone to Iceland to write about snow, but I found that snow had other ideas – it wanted me to do some physical labour. In his poem 'Digging', Seamus Heaney describes how his trade differs from that of his ancestors: 'I've no spade to follow men like them,' he writes, using instead a 'squat pen'. Now I had to put down my pen and borrow a shovel from my elderly neighbour Kristján. Then, of course, I cleared his path too.

Sometimes the drifts were so deep, the white-outs so dense, that Siglufjörður seemed composed entirely of snow. I felt like Marcovaldo, the hero of Italo Calvino's novel of the same name, who

> learned to pile the snow into a compact little wall. If he went on making little walls like that, he could build some streets for himself alone; only he would know where those streets led, and everybody else would be lost there . . . But perhaps by now all the houses had turned to snow, inside and out; a whole city of snow with monuments and spires and trees, a city that could be unmade by shovel and remade in a different way.

Calvino captures how snow makes a familiar place strange; how it can seem to rewrite reality, concealing, clothing, cleansing or suspending the landscape for a time. It *muffles*. It *shrouds*. Like the sheet a magician lays over their assistant before taking out the saw – when it is whisked away, the miracle is not that anything has changed, but rather that everything has stayed the same.

When I first went north of the Arctic Circle in Greenland a decade ago to research glacial ice, I was seeking to escape the distractions of a capital city. I needed white noise. Although my work entails filling pages with words, I have always loved the empty margins the most. There is much poignant art and literature about polar purity and silence, but the longer I spent among the snow, the more I suspected such tropes are born of luxury and distance. It is a view that overwrites the peopled landscape, ignores the tracks of sleds and snowmobiles that cross it, the busy burrows and root systems beneath it. As time passed and I looked more closely, I realised snow does not always appear white. As I listened more carefully, I realised that snow was not silent. I spoke to those who worked with snow, from Inuit hunters to Scottish hill farmers, and noticed that their traditional knowledge was often enshrined in highly differentiated vocabularies.

Fifty Words for Snow is a journey to discover snow in cultures around the world through different languages. The climate is a prism through which to view the human world – just as language can be. It is possible to see back into the distant past and trace the historical movement of people through a single unit of meaning: in Europe, for example, many words (*snow, snee, nieve,* etc.) stem from the same root, the ancient Latin *nix* and Greek *nipha* – the initial *s* comes and goes, without concealing the close connection. Inevitably, a book about climate also looks forward, considering what we miss, as every winter in many countries we see fewer and fewer snowflakes, and some years now, none at all. Just as the ecosystem is changing, so are the languages that describe it and the way they are understood. When I began to learn Greenlandic in 2010 I discovered the fallacy of the many 'Eskimo words for snow', a popular concept that was dismissed by linguists in the 1980s (see the references for further reading on this myth). More pertinently, that same year Greenlandic was added to the UNESCO *Atlas of the World's Languages in Danger*. While many of the languages in this book, such as Spanish and Urdu, can be heard spoken around the globe, others, such as the Inupiaq dialect of Wales, Alaska, are remembered mainly by elders in relatively small communities.

I started to write this book in September 2019 amid debates about Brexit and the climate crisis, while attending Fridays for Future marches in Germany. I finished it six months later, a week before gathering with other masked and silent Black Lives Matter protestors in the UK. The process of tracing a single theme across many languages new to me seemed a powerful way to overcome the borders that were going up around the world. Even under lockdown in a pandemic, it was still possible to voyage around the world through dictionaries. Ironically, one of the first entries I researched concerned the photograph of boys in a snowball fight taken by Robert Capa in war-torn Hankou (modern-day Wuhan) in 1938. Within weeks, the location had become infamous for the outbreak of COVID-19. Meanwhile, I was spending a lot of time in hospitals for another reason. As I began to compile a list of words for snow, my partner suffered a major stroke. My work accompanied me on those anxious autumn months in the ward, to a backdrop of medical equipment hissing and beeping, rather than the soft, reassuring sounds of snowfall. Eventually it became apparent that Anna's stroke had induced severe aphasia; with spring, as some of her words began to return, in fragmentary and often puzzling forms, I grew to appreciate the complexity of language loss anew. The issue of vanishing vocabularies, all too easy to romanticise, was revealed to be heartbreaking – but I appreciated all the more the power that the ability to access even a single word can bring.

The point 'where everything begins' – this is how Traci Brimhall describes the impurity at the heart of the snowflake in her poem about *penitentes* quoted in the epigraph. A snow crystal is part of the endless cycle of the water molecule: from its six-cornered solid state it becomes liquid and then gas, and thus a snowflake that falls on the glaciers of the Rwenzori peaks in Africa might melt and evaporate and later freeze again and fall in the apple orchards of Kashmir, and melt again and fall fifty times and more. Just so, a single unit of meaning – one word for snow – offers an approach to new places, a clear path of understanding to travel forwards along. For language allows us, like Marcovaldo, to unmake our cities and dream them differently.

1.

Seaŋáš

granulated snow
(Sámi)

She picks up her sharpest flint and scratches a few lines in the rock face. One strong horizontal stroke, and another below it, then four long horizontals and, getting into it now, a bit of cross-hatching to fill the space between the others. The flint arcs, almost as if her hand had slipped, and she follows that with a series of staccato chips, each powered with the same flick of her wrist she uses on her drum. Over 14,000 years and at least one ice age later, these confident lines on the wall of a Welsh cave still unmistakably show the image of a beast with magnificent antlers. When the whole world's climate was colder, reindeer roamed across southern Europe and were known in New Mexico, according to the cave drawings left behind by Stone Age tribes and the Clovis people. The flints that drew the animals would also be turned on them.

Today reindeer are creatures of the polar north, living in areas such as Guovdageaidnu in Norway, where snow covers the ground for more than half the year. All through the long winters, during which temperatures can reach as low as minus 30° Celsius, the reindeer graze on the high plateau. They dig down through the snow using their hooves or antlers to find lichens to eat. In spring, lush grasses begin to emerge from the deep snowdrifts on the coast, and it is time for the reindeer to start their great annual migration north to summer feeding grounds by the sea. They are guided by the Sámi people, who have long subsisted in this harsh climate as fishers, trappers and reindeer herders. The spring weather and depth of snow decide when herders begin to move and how fast. They know that cold, crisp ground provides ideal conditions to move their animals swiftly across the plains to the coast. They will often drive the reindeer through the night, waiting for the evening frost to form a light crust on the snow, or *skavvi*, after the sun has thawed the surface during the day. They will rest when the afternoon sun causes *soavli*, or slushy snow. While they are on the move, the reindeer – or at least their traces – are visible on the snow, so that animals might be found again if they go wandering or join other herds.

The Sámi language reflects the herders' intimate relationship with their environment. The rich terminology for snow and ice includes words to describe the way snow falls, where it lies, its depth, density and temperature. One of the most significant types of snow for the Sámi is *seaŋáš*, or loose granulated snow, which forms at the bottom of the snowpack from January to April, a little like 'depth hoar' in the international snow classification. Snow takes on *seaŋáš* consistency during a cold winter, and it improves grazing conditions: it is easy for reindeer to dig through *seaŋáš* to the lichen growing beneath. Since *seaŋáš* melts rapidly, it also provides a vital clean water supply for the travellers. It's not surprising that some Sámi terms for snow relate to its influence on the lives of reindeer, such as the unwelcome state of *moarri*, which is 'the kind of travel surface where frozen snow or ice breaks and cuts the legs of animals'. But while there are around one hundred Sámi terms for snow, the words relating to reindeer are estimated at over a thousand. And yet, to find out what the woman in the cave knew of the reindeer, we have little choice but to let the picture speak.

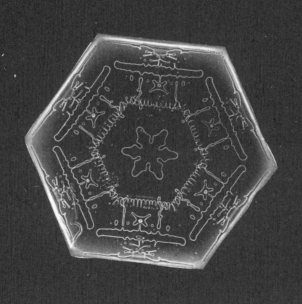

2.

Yuki-onna

snow woman

(Japanese: 雪女)

Taoist philosophy suggests that when there's an abundance of any natural matter, a life will come forth from it: the river will create its own fish when the water is deep enough and the forest will produce birds when the trees are dense enough. And so it follows that a woman may be generated in the heart of a snowdrift.

Nowhere in the world are the drifts as profound as in the mountains of Japan. In the remote highlands of the Japanese Alps, a series of three high mountain ranges – the Hida, Kiso and Akaishi – that bisect the main island of Honshu, the annual snowfall can be as great as 40 metres. The world record for the deepest snow was measured further west, on the slopes of Mount Ibuki in 1927, although it is hard to verify such records when so few of the planet's snowiest peaks are equipped with snow gauges or even accessible to meteorologists and other mortals.

Yet sometimes a mysterious figure can be seen, emerging from a disorienting white-out on the Honshu hills. The *yuki-onna* is among a class of supernatural monsters, spirits and demons (*yōkai*) whose alluring appearance belies their profound menace. The first *yuki-onna* was encountered by a poet in medieval times, her name formed from the words for snow (*yuki* 雪) and woman (*onna* 女). Thousands of sightings have been recorded since, but one constant characteristic is a similarity to snow: her skin is cold; her hair is silver; she dresses in white. She drifts through the hills, her beauty all the more fascinating because it is so fleeting. Many tales of the *yuki-onna* dwell on her swift disappearance – in one story she transforms into a flurry of snowflakes in a puff of wind; in another, she melts away having been persuaded by her lover to take a bath, leaving behind only a few brittle icicles floating in the water.

In these tales, the *yuki-onna*'s desire for human lovers is usually satisfied, but her stay in the mortal world is thwarted by the folly of the people she encounters. One story, about a *yuki-onna* who had a relationship with a woodcutter for many years, was told to the writer Lafcadio Hearn by a neighbour in the district of Musashino. The story was published in the last of Hearn's many books on Japanese culture in 1904, the year he died.

One evening, two woodcutters were coming home from the forest when they were caught in a snowstorm. The wide river they crossed daily was impassable, but luckily old Mosaku and Minokichi, his young apprentice, found shelter in a ferryman's hut. The two men slept, despite the wind blowing outside and the snow beating on the window, but Minokichi was woken in the night by a flutter of snowflakes entering the room. The door seemed to have blown open. By the snow-light (yuki-akari) he was amazed to see a woman dressed in white, bending over Mosaku and blowing upon his face – her breath was like white smoke. But the old man did not stir.

The yuki-onna glided towards Minokichi, and stooped low and then lower over him, until their eyes met and her face almost touched his. She was very beautiful, but the light in her eyes made him afraid. After staring at him for a long while, she said, 'I thought I was going to kill you, the same as that old man, but I will not because you are young and handsome. You must not tell anyone about this incident, Minokichi, even your mother. If you tell anyone about me, I will kill you.'

Minokichi was astonished that she knew his name, and promised he would keep the secret. She turned from him, and left the shelter. Then he sprang up and looked out the door, but he could no longer see the woman nor any tracks in the snow. He called to Mosaku, and was fearful because the old man did not answer. He reached out to touch his body, and found it was cold.

It took Minokichi a year to recover from that terrible night. He picked up his trade again, and went to the forest alone every day. While walking home one afternoon he met a beautiful young woman named O-yuki, and within weeks they were married. Their life together was happy and blessed with ten children, but O-yuki did not seem to age. One evening, when the children were asleep, Minokichi looked at his wife as she sat sewing by a paper lamp. He was moved to tell her about the day he saw Yuki-onna. 'When I see you with the light on your face, I am reminded of a strange thing that happened to me when I was eighteen. I met the most beautiful woman, she was so very like you . . .'

Before he could say more O-yuki stood up and screamed, 'That woman was me! I told you that I would kill you if you ever told anyone about that night. But I'll let you live because of our children. You'd better take very good care of them, or I will treat you as you deserve . . .'

And so the Yuki-onna spared Minokichi's life a second time, but she shed her own human form. After hissing her last words to her partner of so many years, she transformed into a bright white mist – and swiftly spiralled up into the roof-beams, and vanished.

Are all human encounters with the elements so ill-fated? Is it possible to keep our most profound dealings with nature a secret? Will the snows stay forever, or will winter turn to spring?

Whether the *yuki-onna* is a malevolent ghost stealing lonely lives in the wilderness or a supernatural beauty living in disguise among humans, she affirms the transformative qualities of snow.

Today a 90-km road winds through the Japanese Alps southwest of Nagano, known as the Yuki-no-Otani or 'Snow Corridor'. The pass is cleared daily in winter, so that travellers can speed along in coaches and cars, steering between snow walls that reach up to 20 metres high. Thanks to the Yuki-no-Otani there is no longer any risk that people will be trapped in the hills or forced to stay overnight in rickety riverside shelters. Yet the reason this dramatic mountain pass seduces so many travellers is not the comparative safety and accessibility of the route, but the mystery of the miles of deep snow on either side, through which – against almost impossible odds – it carves its way. Some drivers may even long to wander off the track to explore the unmarked snow, in search of their own beautiful *yōkai* and the satisfaction of as yet unknown desires.

3.

Immiaq

melted ice or snow; beer
(Greenlandic)

A few hours into a long-haul flight between North America and Europe the passenger who has a window seat and clear skies can fall into a trance at the snow-covered gneiss mountains vanishing towards the horizon. Zoom out to the map view, where the whole of Greenland is visible from Cape Farewell in the south to Kaffeklubben Island in the north, and the country looks like a giant teardrop rolling from the temperate zones towards the Arctic Ocean. It may be the world's largest island, but most of the surface is uninhabitable. The central region is engulfed by centuries of snowfall, which has compressed to form an ice sheet thousands of metres deep. From this dense mass of ice pitted with blue melt pools, outlet glaciers creep towards the coast. There, with a great roar, they calve into the fjords and the icebergs drift out to sea. The margin where sea and land meet is also covered with a shifting layer of ice that forms gradually in winter and melts away again in spring.

In the distant past, people came to this dynamic coastline from lands to the east and west and settled here, even though survival through the dark, Arctic winter was doubtful. With no means of growing crops on the barren rocks, these homes were a temporary refuge – often no more than a base from which to set off on expeditions to seek food. Communities and hunting grounds were connected by sheltered harbours and trails over the sea ice rather than roads. Hunters journeyed wearing goggles to protect their eyes from the blinding white glare of the snow, seeking the best spots to send down fishing lines or wait with harpoon ready for seals to appear at breathing holes. A hunter had to be self-reliant – carrying all the supplies that might be needed for days on their sled. But there was no need to bring drinking water when everywhere fresh snow and ice could be collected and melted in a pan over the fire. Thus the Greenlandic word *immiaq*, meaning melted ice or snow, also began to refer to drinking water.

In the nineteenth century food imports began to arrive in Greenland at the demand of Danish settlers, and soon people were referring to other drinks as *immiaq* too. Home-brewed beer was drunk by many hunting families. A 1960s recipe for *immiaq* from the far north-west of Greenland recommended mixing

1 kg malt, 50 g hops, some yeast and 2–3 kg granulated sugar – depending on how strong a beer was desired.

Strength is not everything: now the craft-beer revolution has reached the Arctic. In 2012 a small brewery was established in Ilulissat, not far from the terminus of one of Greenland's fastest-moving glaciers. Brewery Immiaq prepares a range of beers, from a dark, silky Christmas beer to a light pink pilsner. Its bottles are raised to the lips of grateful tourists and climate scientists in hotel bars overlooking a sea of icebergs. The great glacier Sermeq Kujalleq may be a UNESCO World Heritage Site, and thus recommended for preservation, but that cannot change the fact it continues to calve around 46 km^3 of ice every year – an amount that would cover the annual water consumption of the USA. This ancient ice is melting fast enough without the aid of any hunter's stove.

4.

Smoor

to perish in a snowdrift
(Scots)

Few professions know the outdoors as well as a shepherd, and those of the Scottish Borders are no exception. Life working on the moors is relentless, especially in the cold, dark months between putting the tups (rams) to the ewes in late November and lambing in the spring. To help ensure the survival of sheep (and shepherd) in the bitter winter, the flocks are moved downhill to less exposed pastures in the valleys. While the rhythm of these seasonal tasks has changed little over the centuries, working conditions do fluctuate; extreme weather events are not a new phenomenon, and the eighteenth century saw terribly cold winters, much harsher than those I endured as a child, just before the millennium, living in a hamlet nestled into a valley in the Cheviot Hills north of the border.

When my friend Morag and I had grown out of sledging and snowball fights (although never of making snow angels), we amused ourselves by gazing at the frosty stars and singing whisky melodies on moonlit midnight walks covering the mile or so to her family's farmhouse, tripping arm-in-arm over potholes and sliding on the unsalted single-track roads that linked our homes. The next morning, long breakfasts turned into lunches around the farmhouse kitchen table, as neighbours came and went, bringing bootleg records and worming pills and taking away home-made haggis, or a few slices of date and walnut cake wrapped in tinfoil. As we dreamed of gap years in Australia and Japan, Morag's father would begin planning his next Shakespeare production: each summer he transformed local shepherds and teachers into actors and the small, straw-lined arena of the local auction mart into a stage. In that bookish farmhouse there was plenty to read on a winter afternoon – whether your taste was for tales of fantasy and adventure such as the Discworld series or classic literature on animal husbandry. While Morag was arguing with her sisters, I roamed the shelves.

Much of the knowledge about historic herding in Scotland comes from James Hogg, the author of *The Shepherd's Guide: Being a Practical Treatise on the Diseases of Sheep* (1807) as well as several novels and collections of poetry. Inspired by the work he did from his small bothy in the Yarrow Valley, he wrote unsentimentally about the environment, with first-hand experience of the catastrophic Scottish weather: 'Whole flocks are sometimes smoored by huge wreaths of snow shooting from the hills upon them.'

Smoored? It wasn't a word I heard often. Hogg glosses it thus: 'Smooring. This is occasioned solely by the shepherd's not having his flocks gathered to proper shelter.' *Smoor* is not exclusive to sheep. A current Scots dictionary defines the verb as: 'To be choked, stifled, suffocated, to suffer or die from want of air, especially to perish by being buried in a snowdrift.' It is also found in the work of one of Hogg's contemporaries (and subject of another of his books), the poet Robert Burns. In his long narrative poem of 1790 'Tam o'Shanter', Burns writes about a drunken farmer, chased by witches. As he flees, he passes another unfortunate person, who has succumbed to the cold:

> *By this time he was cross the ford*
> *Where in the snaw the chapman smoor'd.*

Clearly a useful word for anyone embarking on an unsteady midwinter walk.

The shepherd's duty to find shelter for their flock was no small matter in the eighteenth century. In 1772, for example, snow lay from the middle of December to April, and the sheep were so weak they couldn't move. Hogg details the many dramatic storms that battered Berwickshire, such as that of 24 January 1794, when the wind blew with 'peculiar violence' between Crawford Moor and the border. Seventeen shepherds died and 'one farmer alone, Thomas Beattie, lost 72 scores [of sheep]'. A later book by Hogg, *The Shepherd's Calendar*, opens on 13 February 1823, a dreich day, with 'the snow lying from one to ten feet deep on the hills':

A partial thaw had blackened some spots here and there on the brows of the mountains, and over these the half-starving flocks were scattered, picking up a scanty sustenance, while all the hollow parts, and whole sides of mountains that lay sheltered from the winds on the preceding week, when the great drifts blew, were heaped and over-heaped with immense loads of snow, so that every hill appeared to the farmer to have changed its form. There was a thick white haze on the sky, corresponding exactly with the wan frigid colour of the high mountains, so that in casting one's eye up to the heights, it was not apparent where the limits of the earth ended, and the heavens began.

The Shepherd's Calendar advises landowners on how to create good shelters for their sheep. The best, Hogg wrote, were natural: an enclosure of 'clumps of Scots firs, which, when grown up, keep the flocks safe and warm, though the tempest be ever so fierce'. However, as trees take time to grow, he recommended also building *stells* – the round, stone-walled structures that are so beautiful in their simplicity that contemporary land artists such as Andy Goldsworthy have copied their forms. Hogg's advice:

> The best form in which he can make these is that of a complete circle, or octagon . . . with a door in it, at which the sheep may go in and out. This door should always be made to face that place near by where there is good natural shelter . . . for there is no natural pasture that is not sheltered from some airth. . . . There is no kind of stell . . . so safe as this; sheep are never smoored in them, for the wind whirls the drift around them, and accumulates it in large pointed wreaths on the opposite side . . . the sheep, if once acquainted with them, will come running to them on a cold night from every direction.

The shepherd's dramatic battles with the elements to rescue sheep were not without reward. On returning to their own human shelters, they could expect 'old Janet's best kebbuck, and oatmeal cakes, and preeing the whisky bottle'. There was little need to turn to the glossary to understand 'preeing'.

5.

Hima

snow

(Thai: หิมะ)

It has snowed only once in Thailand – allegedly. On 7 January 1955, in the city of Chiang Rai in the far north of the country, a snow flurry followed rainfall at six in the evening and lay until morning. Or so say some meteorologists. Others claim that the white areas, just visible in grainy photographs in local newspapers, must be hail, not snow. In a tropical region usually characterised by dramatic monsoons, this mild weather event has gained notoriety. Temperatures in the mountains can drop to below freezing at night, causing dew to turn to frost on the leaves, but snow is unknown. Even if the people of Thailand have never awoken to snow in the streets, at least the Thai language has a word for it, allowing everyone to continue to debate whether it appeared that evening – or not.

6.

Kunstschnee

artistic snow

(German)

How often do humans find their emotions mirrored in the weather? Even in make-believe, we have learnt to respond to the tug of a meteorological metaphor. On a stage or film set a winter landscape can usher in a cool mood or evoke chilly emotional endings. But can the subtle nature of snow itself ever be convincingly replicated? The German term *Kunstschnee* perfectly captures the fine art that lies behind winter illusions such as dry ice and fake snow.

Foley artists have a tried-and-tested trick: to create a deep-and-crisp-and-even crunching snow sound effect for radio, cornstarch is packed into a plastic bag, and then the artist rolls their fist over it. Creating visual illusions has proved more challenging. Technicians soon learned that the secret of success was to make snow from materials that refused to melt under the studio lights. Long rehearsals and the Californian climate demanded sparkling drifts that were not dependent on time or temperature. Experiments were made with dustings of gypsum, heaps of bleached cornflakes and scatterings of asbestos; a Styrofoam snowman appeared in the movies as early as 1941 – the film was *Citizen Kane*.

Today one of the main players in the field of *Kunstschnee* is Snow Business. For over twenty years, from locations in England and Germany, the company has wowed the most formidable film directors with ingenious artifice. Their strapline is 'Wir machen Winter' or 'We create winter', and they have around 150 varieties of environmentally friendly *Kunstschnee*.

With *Kunstschnee* a designer can 'dress' a location precisely, with as much depth or density as the plot requires. Snow can be made to fall on cue at any time of day or night, anywhere in the world. Fake flakes can be tinted red to convey the carnage on a Napoleonic battlefield, or grey for the ordure of a Victorian city. Spain can become Spitzbergen, Ealing can look like Everest. Snow Business laid powder snow for the Cold War drama *Bridge of Spies*. It blew over Berlin pavements in *The Bourne Ultimatum*, covered a playground in *The Book Thief* and was sprinkled over the mysterious 'sparrow school' where Jennifer Lawrence learned her trade in *Red Sparrow*. Even before the actors assume their positions, a team of snowmen has set up the illusion.

In *Blade Runner*, a film characterised by a nihilistic urban landscape of decaying skyscrapers and cooling towers, lit by blinking neon screens and car headlamps, extreme weather signals integrity. Remember that epic downpour on a warehouse roof, during which the dying replicant Roy Batty (portrayed by Rutger Hauer, clutching a snow-white dove) delivers his 'tears in the rain' speech? His theme is simple but transcendent: all the dramatic galactic and historical conflicts he has seen will disappear in time, just as the rain washes away his tears. Replicant tears and studio rain combine to suggest Roy Batty has as much soul as any genuine human character in the film.

When it was decided that the denouement of the sequel, *Blade Runner 2049*, would be a snow scene, special effects supervisor Gerd Nefzer called Snow Business. The script has snow falling softly on replicant police officer K (Ryan Gosling) at the moment he meets his end; he opens his hand to the sky and gently catches a tiny snowflake in his palm. 'That's basically a mix of water and washing up liquid,' says Lucien Stephenson, Snow Business CEO. Fittingly for a film about authenticity and its replication, the snow was not real. Snow Business deservedly shared the success when Nefzer won an Oscar for Best Visual Effects. Can we suspend our disbelief when we know that the snow in a movie is fake? Is it any more disappointing to discover a painted studio set, a green screen or an actor feigning love? Now that snow is becoming more of a rarity in winter, *Kunstschnee* is moving from film studios to the ski resorts. Will there be any real snow at all when the year 2049 arrives?

7.

Onaabani Giizis
Popogami Giizis

Hard Crust on the Snow Moon
Broken Snowshoe Moon
(Ojibwemowin)

There are many origin stories for snow: the six-cornered crystals fall in suburban parks in the temperate zones, at high altitudes on remote mountains, and at the extreme ends of the Earth at the poles. Wherever it falls, snow requires two weather conditions: a cold temperature and moisture in the atmosphere. (Even in Antarctica, there is a snowless region known as the Dry Valleys, because although it is very cold, there is not enough humidity.) Humidity is not in short supply over the Great Lakes of the United States, which generate 'lake-effect snow'. As cold, dry air passes over a large body of water, it picks up warm air, which rises and cools, and the moisture condenses to form clouds. The clouds produce heavy snowfall as soon as they pass over land, in this case often downwind in 'snowbelt' states, such as Michigan and New York.

The ancestral lands of the Anishinaabeg people cover a vast territory that includes the waters of Lake Superior. In their calendar, the names for different months describe the natural world and seasonal events during the cycle of days from new moon to new moon. In Ojibwemowin – the language of the Anishinaabeg – moon names vary between dialects, reflecting variations in climate across the region. For those on the west of Lake Superior, *Onaabani Giizis* or 'Hard Crust on the Snow Moon' falls in March, followed by *Popogami Giizis* or 'Broken Snowshoe Moon'. Surprisingly perhaps, the two snow moons appear in late spring. The cold theme returns in November, with the 'Freezing Moon' (*Baashkaakodin Giizis*) and, in between, the moon names celebrate summer flowers and autumn berries, grains and falling leaves.

The names *Onaabani Giizis* and *Popogami Giizis* hint at the traditional knowledge that kept travellers safe on long hunting trips over the snow-covered landscape. This wisdom held that as winter turns to spring, the snow begins to melt during the day and freezes again at night. This process creates a layer on the exposed surface of the snowdrifts that is stronger than the powdery snow below, like the crust of a bread roll. Travellers must judge the thickness of this crust to determine whether it is 'unbreakable' (will it support my weight?) or 'breakable' (will it collapse beneath me?).

These travellers are not alone on their journeys. They might hear a rustle in the pine trees, glimpse a shadow in the woods, or see the tracks of bear, wolf or snowshoe hare vanishing into the distance. The hare is a relatively small mammal, but it leaves behind a significant trail – as its name suggests, it has evolved oversized feet enabling it to move swiftly through deep snow. An ingenious trick, and one that people living in northern regions with extreme winters have adopted too: designing footwear that spreads the weight of the traveller's body so that it does not sink into snowdrifts. Traditional snowshoes are made of ash or other hardwood, which is steamed or soaked to make it pliable, then bent into shape. This frame is laced with rawhide – strips of moose, deer or caribou skin – an intricate latticework that looks like the waxy walls of honeycomb cells. This geometric pattern is not only beautiful but also practical as it prevents the snow clagging to the shoe. Like dialects, snowshoes came in many different shapes, from the circular design of the far north to the Cree snowshoe, which is nearly six feet long and turns up at the toe – perfect for slipping between trees in the forest.

Popogami Giizis marks the end of winter. The snow is melting and it is time to hang up your snowshoes – but before you do so, you'd better make them ready for the next cold season, taking a few hours to mend the wooden frames if they've broken and re-lace the rawhide – perhaps by moonlight.

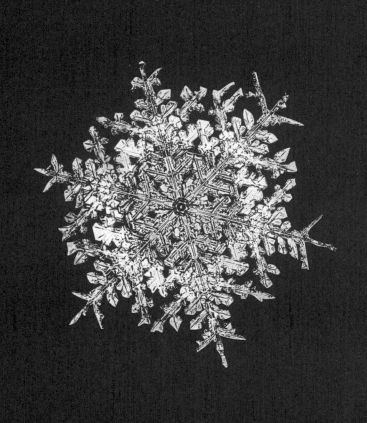

8.

Sheleg

snow

(Hebrew: שֶׁלֶג)

'He giveth snow like wool: he scattereth the hoarfrost like ashes.
He casteth forth his ice like morsels: who can stand before his cold?
He sendeth out his word, and melteth them: he causeth his wind
to blow, and the waters flow.' Psalms 147: 16–18

The psalms praise a God who dispenses natural phenomena
with abundance. Snow is part of the cycle of changing seasons:
impulsive, terrifying and temporal. In ancient times the highest
peaks in the Middle East were white through the year. Snow was
transported from the Lebanon mountain range to the coastal
cities of Tyre and Sidon, and Damascus further inland, for use
in food – a sign of the growing luxury of the Solomonic period.
It was collected and stored in clefts in the hills to cool the drinks

of labourers in the sweltering harvest fields – a practical use that appears as a metaphor in sacred texts: 'As the cold of snow in the time of harvest, so is a faithful messenger to them that send him: for he refresheth the soul of his masters,' wrote the author of Proverbs 25:13.

It is easier to appreciate the refreshing qualities of snow in an arid climate, one that usually experiences extremes of heat, rather than at the ice deserts of the poles. Snow is rarer – indeed almost miraculous – and so it becomes symbolic of holiness and purity. The prophet Job asks (38:22): 'Hast thou entered into the treasures of the snow, hast thou seen the treasures of hail?'

One person who entered into the treasures of the snow in every sense was Hillel the Elder, who was born into a poverty-stricken family in Babylon around 110 BCE. Hillel loved the Torah and travelled to Israel to study it, but there his entrance fee to the Torah study hall cost a quarter dinar – half what he could earn in a day's work. Yet he was so pious that, even in poverty, he never considered spending his money on anything else.

One winter day, Hillel could find no work and was unable to pay his fee to hear the teachings of the rabbinic sages Shmaya and Avtalyon. The Talmud records how he was so determined to continue his learning that he decided to scramble onto the roof of the study hall, where he could lie beside a skylight to eavesdrop on the discussions within. It was a cold evening, and soon snow began to fall. Hillel remained on the roof all night, mesmerised by the Torah teachings, and gradually lost consciousness as his body was covered by the flakes. The next morning, Shmaya realised that there was a figure blocking the light from above. He sent his students up to the roof and, even though it was Shabbat and the lighting of flames was forbidden, they kindled a fire to warm the frozen man – saving a life takes priority. Hillel was revived from out of the snow and continued his studies – in time becoming one of the most influential rabbis in history, the spiritual leader of the Jewish people, and founding a school that bore his own name. He was known as a humble and patient man, who exhorted his students to love their fellow humans, to treat their bodies (the house of the soul) with respect, and never to procrastinate – lessons that he may have drawn from his intense experience of the snow, as well as from his elders' teachings.

9.

Sastrugi

sharp ridges on the snow
(Russian: застругu)

The vast icy plateaus of the polar regions might appear to be expanses empty of any recognisable features, but it is possible to practise the art of pathfinding even here. Polar explorers learn to look for reflections in the sky to assess whether there is water or ice ahead, or to orient their steps by observing the direction of sun and shadows. Extreme weather leaves its traces on one infinitesimal snow crystal after another, and explorers can take bearings by reading the forms created by the sweep of wind over snow, such as snow mounds, ripples and *sastrugi*.

Under the influence of a steady wind, the kind that whips inland from the sea across the drift ice, snow particles accumulate like the grains in sand dunes. Over time these drifts that have been built up by the wind are eroded by it in turn, and sculpted into new shapes. When the winds slacken, the atoms in these beautiful pillars of snow consolidate by the processes of sublimation and recrystallisation and the forms harden to ice. Like cloud-watchers, an onlooker mesmerised by *sastrugi* might find their shapes evoking other wonderful forms in the imagination.

The word *sastruga* comes from *zastruga* – 'groove' or 'small ridge' in the Siberian dialect of Russian – and was first used of ice in Germany during the nineteenth century (*sastrugi* and *zastrugi* are plurals). Now a term used universally among ice experts, it has moved back into usage in the wider Russian language, a reminder that words – like those who use them – can travel home as well as out into the world.

The coast of Russia accounts for over half of the Arctic Ocean's margin, and the nation has long experience in the navigation of polar waters. In the Southern Hemisphere, from a research base at Vostok, Russian scientists are drilling deep through the ice to explore the ancient lakes of Antarctica. No wonder then that Russian is one of the four languages included in the World Meteorological Organization's glossary *Sea Ice Nomenclature* (the others being French, Spanish and English). This indispensable guide for navigators, which can be found in the bridge of every icebreaker, defines *sastrugi* as 'sharp, irregular ridges formed on a snow surface by wind erosion and deposition'.

The tribulations of travelling across these sharp, exposed ridges are a well-worn theme in polar explorers' journals. The hardest and most resilient *sastrugi* force skis up and down and from side to side. Ernest Shackleton encountered these 'annoying obstacles to sledging' on an ascent of Mount Erebus during his 1907–9 expedition. His fellow explorers, who had just pulled a sledge on their hands and knees up a steep glacier, were dismayed to be confronted by that 'which impeded their progress somewhat'. In his account of this epic journey Shackleton implies the cursing that tackling this icy ridge-and-furrow system entailed. 'The sledgers', he writes, 'had much trouble in keeping their feet, and the usual equanimity of some of the men was disturbed, their remarks upon the subject of sastrugi being audible above the soft pad of the finnesko, the scrunch of the ski boots, and the gentle sawing sound of the sledge-runners on the soft snow.'

The *sastrugi* might have caused frustration, but they served a purpose too. The explorer crossing drift ice can navigate by *sastrugi*, since the ridges run consistently parallel to the direction of the wind at the time they are formed, and are less temporary than many other snow features. Shackleton noted this to be especially useful 'when the sky is overcast and the low stratus cloud obliterates all landmarks. At these times a dull grey light is over everything, and it is impossible to see the way to steer unless one takes the line of sastrugi and notes the angle it makes with the compass course, the compass for the moment being placed on the snow to obtain the direction.' The very qualities that make a journey most arduous can be a means to find a way forward.

10.

Hundslappadrífa

snowflakes big as a dog's paw
(Icelandic)

One of the first Icelandic tourists, P. C. Headley, wrote of riding across the island on horseback 'in a sort of grave exhilaration, gazing as in a dream at the hills, and drinking in the sunlight, content with silence and the present'. More and more people have been journeying to Iceland, hoping to find silence away from the modern world in the great snowfields and frozen tongues of glacial ice. Yet the glaciers are vanishing. The first to melt away was Ok, in 2019, and scientists fear that of 400-plus glaciers, none will remain by 2200. The barely perceptible crackle of frost in the air has been replaced by throbbing bass, and the howling wind by elfin, high-pitched vocals. Today Iceland is as well known for its music as the ice from which it takes its name. In the 1990s the enigmatic singer Björk exploded onto the indie music scene, introducing many English native speakers to the concept of diacritics – and then came a riot of Icelandic indie and post-rock bands, of which Sigur Rós remains the headliner. While Björk tactically chose to sing in English, Sigur Rós achieved success even with British and American audiences who could not understand the words.

Over time Sigur Rós lyrics became impossible for Icelanders to understand too – some of the band's songs aren't sung in Icelandic either. The lead singer Jónsi began to apply his eerie falsetto to a newly invented language: 'Hopelandic', or *vonlenska* in Icelandic. This is not so much a language (since it has no vocabulary or grammar), but rather 'a form of gibberish vocals' that fits to the music and acts as another instrument. Many singers do this while composing, when they've decided on the melody but haven't yet written the lyrics, but the exercise rarely makes it into the final score. Hopelandic appears on the Sigur Rós albums *Von*, *Ágætis byrjun* and *Takk...*, and throughout the enigmatic album *()*. The name is suggestive – is hope beyond words?

It was probably only a matter of time before the human voice melted away from Jónsi's productions altogether. In 2019 he created a haunting ambient album, *Lost and Found*, with Alex Somers, which the pair dropped on the world just before a North American tour. 'It falls somewhere between what was, what is, and what will be,' Somers said, cryptically. 'Sound friends that you may have heard or seen before appear; familiar, but different.' One track is named 'Hundslappadrífa', which translates as 'snow as big as a dog's paws'. These large snowflakes cluster together, and fall swiftly and softly from clouds in calm weather, softening the bleak landscape. Children are glad when *hundslappadrífa* blankets the grey city streets as it is ideal for making snowballs. According to Icelandic meteorologist Trausti Jónsson, the term first appeared in print in a newspaper report in 1898. Iceland has a rich vocabulary for snow: the default is *snjór* (snow), while *mjöll* is snow that has just fallen; there are variations like *skæðadrífa* (bright snow) and *logndrífa* (calm snow). But *hundslappadrífa* has a particular magic of its own.

Music has often been used to evoke snow soundscapes, from Tchaikovsky's twinkling waltz of the snowflakes in *The Nutcracker* (characterised by crystalline percussion: an insistent triangle, a fluttering tambourine) to Marvin Gaye's hit 'Purple Snowflakes', with its hallucinogenic promise of purple snowflakes falling from blue skies. Yet music and snow are not natural bedfellows: one of the markers of a fresh snowfall is quietness, since air pockets in snow absorb sound. The song 'Hundslappadrífa' elides the experience of precipitation and listening: musical notes and flakes falling, chords gradually accreting in the listener's memory. In the end this delicate, otherworldly orchestra fades to a soft and dusty patter, which can only be a real field recording of snowfall. The microphone is held up into the swirling flakes, documenting them as other artists have angled a camera lens down upon them – and then there is silence.

11.

Sheen

snow

(Kashmiri: शीन्*)*

Water is abundantly present in the landscape of Jammu and Kashmir, a state that boasts such beautiful scenery that it is widely known as 'Heaven on Earth'. Water in its solid state can be seen on the peaks of the mountains that form the westernmost ranges of the Himalayas, and in the glaciers on their slopes, the forerunners of which once carved out the fertile valleys below. Its liquid state is ever present in the torrents of meltwater that race down from the watershed, gathering as they go into sparkling ice-cold rivers and waterfalls. The water vapour rises again, forming clouds that in turn become rain, refreshing the fields and orchards; this particular paradise is known for its apple industry.

In the 'Apple Basket', the crops are ready for harvest in October, but many farmers wait until November as apples fetch a higher price around Diwali, the festival of lights. Every orchard owner makes their own decision about how long to keep fruit on the trees. They are aware that cold weather can damage the crops, but heavy snow doesn't usually fall until December, by which time all the fruit has been picked and sent to market. But the farming calendar is having to adapt to an increasingly unstable and unpredictable climate. The valley has seen droughts and floods, and years without snow as well as heavy snowfall.

One November evening in 2019, snow fell swiftly overnight, burying the orchards in deep drifts. According to the weather station in Srinagar, rain and snowfall that November reached 118 mm, the highest it had been since 1980. That single snowfall damaged around half the state's 7 million apple trees. Newsfeeds were buzzing with reports of the snow, or शीन् in Kashmiri – a language spoken by (coincidentally) an estimated 7 million people in the state. The apples that had fallen to the ground, or were waiting to be packed away, were buried in drifts; their bright red skins developed dark blotches, ice crystals forming inside the protoplasm of their cells. Mature trees collapsed as the weight of the snow split the gnarled trunks; branches broke away; other trees were completely uprooted; newly planted saplings destroyed. As well as the visible destruction, frost worked its way deep into the trees' woody tissue and xylem cells. As the wood oxidised, it grew discoloured and was soon invaded by wood-rotting organisms.

The loss of crops meant a harsh year for many farmers, and to compound the disaster, they knew how much time the devastated orchards would take to rebuild. Five years for an apple sapling to begin to bear fruit; at least a decade before it matures and yields fully. *Wealth comes like the falling snow, wealth goes like the melting snow*, an ancient Kashmiri proverb has it: 'Yiwawani daulat pēwawún shín; Tsalawani daulat, galawún shín'. Riches accrue slowly, yet they may disappear in a moment.

12.

Cheotnun

first snow
(Korean: 첫눈)

The word for snow in Korean, *nun*, is the same word as is used for 'eye'. And so if you experience the first snowfall of the year – *cheotnun* – with someone you have eyes for, it is said that true love will drift into your arms.

13.

Penitentes

penitent-shaped snow
(Spanish)

At high altitude and in bright conditions an expanse of snow can be transformed into an eerie forest of attenuated blades the height of a human being. The slim shining forms stand close together, the tip of each pointing in the general direction of the noonday sun, giving the masses an orderly, regimented appearance. Out of the corner of one eye, they can seem more like ghosts than glaciers, and so it is fitting that their name recalls the tall, white-pointed hoods worn by brothers of religious orders in the processions of penance during Spanish Holy Week.

The first recorded sighting of *penitentes* was described by Charles Darwin, on 22 March 1835, as he crossed a snowfield near the Piuquenes Pass in the Andes, on the way from Santiago de Chile to the Argentine city of Mendoza. 'These frozen masses,' he wrote, 'during the process of thawing, had in some parts been converted into pinnacles or columns, which, as they were high and close together, made it difficult for the cargo mules to pass.' Darwin believed the formations were caused by wind, like *sastrugi*, which are also known to hinder travellers.

In fact, it is now known that *penitentes* are caused by the sun's radiation, and that their formation draws on the innate character of the snow itself. An expanse of snow is not entirely flat: there are small dips and dimples in the surface. On a bright, dry day the snow in each dip will concentrate the sun's radiation, heating more quickly than the snow around it. Little by little the snow in these dips evaporates. Once a hollow has begun to form, it attracts much more solar radiation than the surrounding snow. Eventually the hollow grows so deep that the snow on which it rests entirely disappears, leaving only the sharp pointed peaks alongside.

The formation of *penitentes* relies on the water molecule in its most mysterious process of transition: ablation, the passage from a solid to a vapour state without taking liquid form. Ablation is a form of sublimation, which in alchemy was considered one of the twelve core processes. A substance was heated to a vapour, and immediately collected as sediment on the neck of the glass alembic. Separation through sublimation was thought to be ruled by Libra (♎), the astrological sign that regulates the balance of the seasons and the length of days.

The *penitentes* are formed from deep snowdrifts, the kind that have fallen day after day, perhaps year after year, the compression of layers of storm memory. It is only through the disappearance of the surrounding snow that the shape of each *penitente* can evolve: the vanished vapour is their lost twin. Nor are the *penitentes* themselves destined for longevity. They are in the league of sundogs, *fata morgana* and other mirages once experienced by polar explorers: objects forged from the equivocal play of light and water vapour, visions that disappear in air as fast as ice melts. What they are making atonement for, nobody knows.

14.

Cīruļputenis

'a blizzard of skylarks'
(Latvian)

Skylarks are small brown birds whose arrival in Latvia to breed in late February and March heralds the spring. The female skylark builds a nest on open ground, and lines it with soft grasses, while the male courts her from the air. His song, described by the poet George Meredith as a 'silver chain of sound / of many links without a break / in chirrup, whistle, slur and shake', is sometimes delivered from so high in the sky that the bird is visible only as a distant speck to the listener below. Female birds seem to prefer to mate with the males that can hover and sing for the longest time.

A 'blizzard of skylarks' is used to evoke the enchantment of a surprise snowfall in springtime – whether the snowflakes fall to the ground as deliriously light and silver as the notes of the skylark's song, or beat the air as powerfully as their wings.

15.

Unatsi

snow

(Cherokee: ᎤᏴᏟ)

From the New Jersey Pine Barrens to the Okefenokee Wilderness, Georgia, many acres of open longleaf pine woodlands march across the eastern United States. As the seasons pass, the appearance of these forests barely changes: the needles turn a shade greyer as photosynthesis slows to a halt in late autumn; in winter a dusting of snow lies along the branches and conceals the pinecones on the ground. Why do pine trees keep their leaves year after year, while other trees let them fall?

The moisture content in a leaf will freeze in winter, just as any water does when exposed to freezing temperatures. This irreversibly damages the plant cells. However, the pine's slim needles have a smaller surface area than other, broader leaves, such as those of the oak and maple, and so the likelihood of damage is much less. The waxy layer that covers the needles also offers protection. As cold weather approaches, the water within the pine needle moves to spaces between the cells and the tree produces a protein or resin that acts like antifreeze, binding ice crystals and causing them to form hexagonal shapes, which are kinder to the leaf. Thanks to this system, evergreen needles can survive the cold through several winters. They fall from the tree only with age and are quickly replaced.

The origin of this ingenious system is explained in a tale attributed to the Cherokee, one of the indigenous people of the Southeastern Woodlands of the United States:

Long ago in the forest, birds, animals and trees could all communicate freely with each other, and different species supported one another in difficult times. Then, as now, when the days grew shorter and the nights grew longer, the smallest birds flew away south to find warmer regions to spend the winter. One autumn, a sparrow who had broken one of his wings could not fly with the other sparrows. When the snow came he was suddenly alone – weak, and scared, and shivering as frost began to bite through his feathers. He hopped from one tree to another, struggling to find shelter from the wind and snow. First he asked the Oak for permission to nest, next the Maple, then the Elm and Aspen. None of these trees wanted to cradle a wounded sparrow in their canopies. One by one, their branches shook the bird back into the snowstorm.

There was only the tall Pine left. The great tree heard a bird sobbing beneath his sweet-smelling branches, and kindly asked what the trouble was. On hearing the sparrow's story, the Pine said: 'My leaves do not offer much protection, they are more like needles, and my branches are sticky with sap, but you are welcome to share what rough shelter I can offer.' And so the sparrow stayed with the Pine all through the winter. He snuggled in to the crevices in the scaly bark, while the wind blew around the tree tops and snow gathered along the branches. During the long, dark nights, he tucked his beak under his broken wing and slept.

At last spring came. The Pine began to put forth the bright new shoots that people call candles, and the other birds returned in flocks. The sparrow leapt from branch to branch with joy, and in doing so discovered that his wing had healed.

When the spirit of the forest heard this story he called together all the trees and admonished those who had not sheltered the sparrow. He punished the Maple, Elm and Aspen by causing their leaves to fall every autumn, leaving them bare and unprotected when the weather was coldest. But the Pine was rewarded for his generosity with the power to remain green all year round.

16.

Theluji

snow

(Swahili)

Colours have great symbolic value to the Maasai, a people whose territories reach from Kenya's Rift Valley plains to the lowlands of northern Tanzania. Black is particularly holy, reminiscent of dark rain clouds, and black clothes were once worn by women as a sign of fertility. The pale colour of cumulus clouds, of snow, and nourishing milk from grazing animals is a sign of peace.

The black and the white merge on the peaks of Mount Kilimanjaro. The vast mountain is composed of three volcanic cones – Kibo ('white mountain'), Mawenzi ('black mountain') and Shira. Mawenzi and Shira are extinct, but Kibo is only sleeping and may erupt again one day. Kibo's peak is often hidden by mist and clouds, but when the skies clear the reason for the name is evident: the summit is capped by massive glaciers formed by centuries of snowfall, shimmering against the blue skies like an optical illusion.

This is one of the few places it is possible to see snow (*theluji* in Tanzania's national language, Swahili) in equatorial Africa. Tourists pay thousands of dollars to hike to the summit of Kibo. They follow in the footsteps of nineteenth-century German explorers, who made the first recorded ascents. (The dominant histories of mountaineering in the Great Lakes region have been based on the records left by colonists from Germany, and subsequently Britain, and they read as part of a wider narrative of conquest.) Yet, in this instance, it is likely these mountaineers were indeed the first to make the climb. The local people (including the Maasai, many of whom today work as guides and porters) chose not to scale Kilimanjaro out of respect for the nature spirits and ancestral spirits who live there. As today's tourists climb up, torrential waters flow down from the summit. Many Maasai villages get their water from the Kilimanjaro glaciers, and so their rapid melting is a matter of grave concern. Here, three degrees south of the equator, drought is always a danger. It can take a village over a year to recover from one, and if drought follows on from drought – if there are only showers, not steady rain, and no other source of water – recovery can be impossible.

Respecting the balance of nature is central to Maasai religion. The Maasai worship Eng'ai, the divine principle that created all life on Earth. According to proverbs, while Eng'ai lived in the sky and was at one with it (Eng'ai means both 'rain' and 'sky'), the divinity also had a close relationship with *enkop*, the Earth. Together, Eng'ai and humans, the forces of Sky and Earth, worked in harmony to create and nurture life. All natural phenomena could be seen as expressions of Eng'ai's divine power and judgement, especially those concerned with the weather and thus coming from the sky – rain as blessing, drought as punishment, thunder and lightning as anger, and rainbows as joyful approval. Kilimanjaro's snowy peak is a meeting point between the human and the divine.

17.

Avalanche

a fall or slide of snow mass on a mountain slope
(French)

When snow falls, we expect it to stay where it lies – if it lies at all. But high in the mountains, every snow crystal is restless – and dangerous. The snowpack is complex and constantly changing; conditions might alter rapidly; nothing can be taken for granted. Despite this unpredictability, ski resorts must guarantee safe slopes for thousands of holidaymakers every day.

The Alps cross national borders – France, Italy, Switzerland, Austria – and the origins of the term *avalanche* reflect this. The word is derived from the Alpine dialect word *lavanche*, influenced by the Old French word *avaler* ('to descend') and perhaps from a pre-Latin Alpine language (the suffix *anca* suggests Ligurian). The word was being used by English speakers from the late eighteenth century onwards, as more people travelled on the Grand Tour to witness for themselves the wonders and perils of this most scenic part of Europe.

In the 'Winter of Terror' (1950–51) there were over 600 avalanches across the Alps, in which 265 people lost their lives. Today, avalanches cause between 150 and 200 fatalities worldwide each year. Vital research into avalanche dynamics takes place at experimental test sites, like Col du Lautaret in the French Alps, where the complex interaction between snowpack, weather conditions and terrain is studied. To help predict the chances of avalanche, scientists study avalanche paths, and dig snow pits and examine the composition of the different snow layers. The size and shape of the snowflakes in each layer are examined for clues.

Scientists have discovered that snow formed from large, loose crystals that have not compacted is weak because there are fewer points at which the crystals connect. The 'powder-snow avalanche' travels at high speeds (up to 330 km/h) accompanied by a wind that can be almost as damaging as the force of the snow. Alternatively, during the spring, the snow on sunny mountainsides will rapidly warm and melt, leading to a slide known as a 'wet-snow avalanche'. The commonest kind of avalanche falls from cool mountain slopes shaded from direct sun, which prevents bonding between snow layers. A fracture in this kind of snowpack ('slab avalanche') spreads swiftly when the snowpack is disturbed by a skier or snow vehicles. Contrary to popular belief, avalanches are not triggered by noise.

All ski resorts in France have an avalanche control programme. Nationwide more than 2,500 ski patrollers (*les pisteurs*) monitor 140 observation posts. Twice a day, they collect information about the wind (its speed and direction) and the snow (its thickness, quality and total height). They measure the snowpack with probes to assess its stability. This data is used to forecast the likelihood of avalanches, and decide where intervention is necessary.

One approach to avalanche prevention is to trigger relatively small flows of snow in a controlled manner. The *pisteurs* use various methods, of which the most basic is also the most risky: they carry explosives to the slopes and position them on the snow by hand, before retreating and detonating them remotely. Today other, safer methods are favoured, which keep the *pisteurs* far from danger. These methods include throwing explosives down from hovering helicopters, or from a rudimentary ski-lift. Some recent inventions sound worthy of a Bond movie showdown: 'Avalex' – sophisticated hydrogen- and oxygen-filled balloons – or state-of-the-art rockets called *avalancheurs*, which precision-fire arrows carrying an explosive mixture of pressurised nitrogen at the slopes. A gas explosion is the least dangerous approach: an incendiary mixture of oxygen and propane gas is held in chambers permanently installed in the mountainside, then detonated remotely from an office, creating dramatic shockwaves. Despite the brave work of *les pisteurs*, every skier should still be aware of avalanche dangers, and precautions taken before venturing off-piste.

18.

Tykky

thick snow and frost that accumulates on
tree branches and other structures
(Finnish)

J. R. R. Tolkien described his first encounter with the Finnish
language, in a letter to poet W. H. Auden on 7 June 1953, as
'like discovering a complete wine-cellar filled with bottles of an
amazing wine of a kind and flavour never tasted before. It quite
intoxicated me.' Certainly Finnish resembles few other European
languages; it is a member of the Finnic family, along with Estonian
and other languages spoken around the Baltic Sea. As a linguist
Tolkien would have recognised that some words, such as *tykky*,
were borrowed from a North Germanic source (it is compar-
able to Old Norse *þykkr* – 'thick, bulky', from Proto-Germanic
þekuz). The unfamiliar, complex forms of Finnish inspired one of
the fictional languages he constructed for his fantastical realm

of Middle-earth: Quenya, a highly revered language spoken by the Elves.

A sighting of the strange, ethereal *tykky* or *tykkylumi* (*lumi* is the word for snow) can be just as intoxicating as a new language or dusty racks of a rare vintage. *Tykkylumi* accumulates throughout the winter in the cold northern regions of Finland. The rime forms when minuscule water droplets in fog, clouds or humid air make contact with the bark of a tree – freezing instantly into a crisp, white coating of crystals. If the wind is strong and blowing from only one direction, only one side of the tree will be covered in rime. Sometimes heavy *tykkylumi* covers the tree entirely: a tall spruce can hold up to three tonnes of snow, and trees have adapted by growing shorter branches so that they do not break under the load. The forest clears itself of snow a couple of times each winter, whenever temperatures creep above zero and there's moderate wind for a day or two. But *tykkylumi* soon starts to build up again, and generally lasts until March, when daytime temperatures rise.

Trees have always been important in Finnish culture. There's even a character known as 'the architect of the forests' in the national epic poem, the *Kalevala*. At the beginning of the world this legendary figure, Sampsa Pellervoinen, sows vegetation over the Earth, creating all the forests, swamps and meadows. Anyone who has wandered in these northern forests in summer, relishing the rich scents of earth and sap, will know they still possess a mythical power. This is heightened when they are viewed against blue and pink skies after a storm has passed, and with snow concealing the evergreens' dark branches and sharp needles. Often *tykky* trees are so heavily matted with snow that their whole shape is changed, bowing with the weight of something that is perceived as weightless. Snow accumulates in fubsy clumps on the branches and the slender crowns slump downward, so the trees create enigmatic shapes, which appear almost on the verge of becoming animate in the short midwinter days and long dawns and dusks. The soft, round forms recall the shape of kuksa cups, made by the Sámi people from the burls on birch trees, or even the bulbous noses and plump white furry bellies of Tove Jansson's Moomins.

19.

Barfānī chītā

snow leopard

(برفانی چیتا :*Urdu*)

A hiker in the rugged wilderness of the Hindu Kush and Karakoram mountains of northern Pakistan might occasionally come across the tracks of another creature in the snow. The world's biggest and most elusive cat is perfectly at home in this cold, arid habitat. The snow leopard roams over ranges of up to one thousand square kilometres at dizzying altitudes in the Himalayan and Central Asian region; the animals take no notice of national boundaries. The names for this top predator are as many as the languages spoken across the region, but they often have in common a reference to snow. Russian has *snezhnyy bars* (снежный барс, 'snowy leopard'), Dari is *palang-e barfi* (پلنگ برفی, 'snowy tiger') and Sanskrit and Hindi *him tendua* (हिम तेन्दुआ,

'snowy leopard'). Likewise, in Urdu, the national language of Pakistan, the name embraces snow *baraf* (برف).

Humans are most likely to sight the solitary *barfānī chītā* as it crosses a snow field, especially at dawn and dusk, or perhaps they might discover a large paw print in the snow after the leopard has passed. A snowy landscape is where the big cat is most visible, but this is not its usual backdrop. The animal lives and hunts among the shadowy clefts of the yellow and grey cliffs, and on rocky mountain passes softened by scrub and sparse areas of grass. Its sandy-white fur, mottled with beautiful dark grey rosettes, blends in so well with the surrounding stones that it is familiarly known as the 'Ghost of the Mountain'. This camouflage is part of the snow leopard's hunting strategy, which is based on secrecy and surprise. Gently, silently and with supreme confidence the leopard stalks the blue sheep and the Himalayan ibex over precipitous rocks, balancing with the aid of its tail, which is the largest of any big cat. When the leopard gets close enough, it will pounce – sometimes by leaping seven times its own body length. The nimble prey do not see their nemesis until it is too late.

It is no surprise, then, that even trained observers might not spot the snow leopard – and sightings are becoming rarer. This iconic species has been assessed as 'vulnerable', and fewer than 3,500 animals are believed to exist in the wild. In Pakistan the population is estimated at a mere 200 to 420 animals, according to the Snow Leopard Trust. The leopard is threatened by poachers, and its habitat is vulnerable to climate change, and the development of roads, railways and hydro-dams. In 2014 the twelve Asian countries that are home to the big cat – Afghanistan, Bhutan, China, India, Kazakhstan, Kyrgyzstan, Mongolia, Nepal, Pakistan, Russia, Tajikistan and Uzbekistan – declared 23 October to be International Snow Leopard Day, and agreed to work together in their conservation efforts. The species that crosses all these borders has created a bond between international communities – and now tech giants are getting in on the act too. The snow leopard population is so hard to catch a glimpse of that it was proving hard to save. To gather data in the past conservationists had to trek to remote regions, and then wait hopefully for a snow leopard sighting; more recently, strategically placed cameras made it possible to observe areas where the snow leopard population was known to hunt, on an ongoing basis. However, it still took scientists days to scan through the many thousands of images that had been captured to find those relevant to their research – as in, those containing a snow leopard, rather than a surprised ibex, or even nothing at all. Now Microsoft has provided sophisticated AI that can scan all the images in minutes, and it can even match up the distinctive patterns on each leopard's side, so scientists can map the journeys of individual animals around their vast territory.

20.

Snemand

snowman

(Danish)

'The Snow Queen' is one of Hans Christian Andersen's best-known and most lyrical fairy tales. The cool monarch, aloof in her ice palace, was introduced to a new generation through its adaptation as the Disney animation *Frozen*. In snow the Danish author found a perfect metaphor for the poignant, ephemeral nature of human life, as well as its beauty. It appears in other stories of his too, such as the tale of a snowdrop (*vintergæk*), which pokes its head above the soil too soon, but finds immortality pressed between the pages of a book of poems. And in 'The Snow Man', Andersen writes about cold weather from the perspective of a creation who depends on it more than any other. The tale opens by evoking the beauty of winter in the countryside:

Towards morning, a thick fog covered the whole country round, and a keen wind arose, so that the cold seemed to freeze one's bones; but when the sun rose, the sight was splendid. Trees and bushes were covered with hoar frost, and looked like a forest of white coral; while on every twig glittered frozen dew-drops. The many delicate forms concealed in summer by luxuriant foliage, were now clearly defined, and looked like glittering lace-work. From every twig glistened a white radiance. The birch, waving in the wind, looked full of life, like trees in summer; and its appearance was wondrously beautiful. And where the sun shone, how everything glittered and sparkled, as if diamond dust had been strewn about; while the snowy carpet of the earth appeared as if covered with diamonds, from which countless lights gleamed, whiter than even the snow itself.

The Snow Man is not described in such enchanting terms. To a soundtrack of 'the joyous shouts of boys, the jingling of sleigh-bells', he is created from the cast-off oddments of human life. For eyes he has two triangular pieces of tile in his head; for a mouth, a broken rake. Hastily made up of random parts on a winter's afternoon, he knows no other season, nor does he understand that he never will live to see another. Luckily, there is an old dog in the yard who can gently steer this naive heap of H_2O towards self-knowledge. The dog has seen new snowmen come and go every winter, and recognises that the tingling in his bones that signals a change in the weather will also be the end of this Snow Man.

The Snow Man soon learns about the mutable solar system. 'How that great red thing up there is staring at me!' he says. He watches the sun rise, and progress across the heavens, and it makes him restless: 'If I only knew how to manage to move away from this place. If I could, I would slide along yonder on the ice, as I have seen the boys do; but I don't understand how; I don't even know how to run.' The old dog warns him that the bright, swift planet is not his friend: 'I saw the Sun, last winter, make your predecessor run, and his predecessor before him. Away, away, they all have to go.'

The Snow Man is admired by two lovers, who joke that he will not be here come the summer. The Snow Man is a novelty to them; by contrast the old dog finds that he has outlived his welcome. Once, he recalls, his owners thought he was handsome; they 'used to kiss my nose, and wipe my paws with an embroidered handkerchief . . . I had my own cushion, and there was a stove—it is the finest thing in the world at this season of the year.'

Much of the charm of the story lies in the Snow Man's child-like discovery of everyday objects – the sun, the lovers, a stove . . . He is especially curious about the stove, which is just visible through a window. The sight gives the Snow Man a new sensation: 'What a strange crackling I feel within me!' He yearns to enter the house to get closer to the stove, but the dog warns him off. 'If you approach the stove, you'll melt away, away.'

The Snow Man doesn't care. He watches the stove, and in the twilight the scene becomes still more inviting, for the stove glows gently, not like the sun or the moon. The light of the flames falls directly on his face and breast, and he can barely endure its beauty. It is a fatal attraction to something that will destroy him. In his stove-sickness, he forgets to appreciate the qualities of the cold. After a long night of ardent stove-gazing, the windowpanes are covered with beautiful ice-flowers – but all the Snow Man notices is that they conceal the stove.

'Maybe it's wrong when we remember breakthroughs to our own being as something that occurs in discrete, extraordinary moments,' wrote another Danish author over a century later, in a novel that became an immediate classic, *Miss Smilla's Feeling for Snow*. 'Maybe falling in love, the piercing knowledge that we ourselves will someday die, and the love of snow are in reality not some sudden events; maybe they were always present. Maybe they never completely vanish, either.'

When the thaw comes, the Snow Man's decline is swift. 'He said nothing and made no complaint, which is a sure sign,' Andersen writes. One morning he melts away completely, and where he stood, a pole remains sticking up in the ground. It is the armature around which the boys had built him up. Now the dog understands his obsession: 'Why, there's the shovel that is used for cleaning out the stove.' The Snow Man had a stove scraper in his body; that moved him to recognise the fire as a kindred spirit.

The figure of the snowman has had a rocky ride in popular culture since Andersen's time – let's skate over the 1980s horror movies – but the same quality of otherness that made the snowman a prototype of evil also made it easy to set him up for a bathetic romance. The pathos in Andersen's stories is laced with the ludicrous. The Snow Man is a kind of Fool, like a pale, silent Buster Keaton, but his motion is all on the elemental level, as molecules change from solid to liquid, rather than in leaping off trains and through windows.

A psychoanalyst could write a long report on the 'Snow Man' complex. Many critics believe that Andersen's life forced his escape into fairy tales; the Snow Man's suffering might have been inspired by his unrequited love for both men and women. Andersen might not have been lucky in love, but appropriately for one who wrote with such empathy of a motionless snow monument, he is commemorated by many statues. His likeness can be found (in more enduring materials) in many places in America – from Central Park in New York City on the east coast to Solvang (a city founded by Danish immigrants) in California. It's also in Japanese theme parks, and of course around Denmark. In Andersen's birthplace Odense, a bronze statue by Jens Galschiøt is sunk in the harbour – as if the author was setting off in search of the little mermaid, or sinking into his own meltwater. When the barometer drops below freezing, the figure appears up to his shoulders in ice.

21.

Mávro chióni

black snow
(Greek: μαύρο χιόνι)

'To appreciate the beauty of a snowflake, it is necessary to stand out in the cold,' wrote Aristotle. The philosopher knew something only too familiar to today's cryologists: that an understanding of nature entailed hard work, often in tough conditions. Aristotle went on to describe not only snow's beauty but also its essence in his treatise *Meteorology*: 'Snow and hoar-frost are one and the same thing, and so are rain and dew . . .when cloud freezes there is snow, when vapour freezes there is hoar-frost. Hence snow is a sign of a cold season or country. For a great deal of heat is still present and unless the cold were overpowering it the cloud would not freeze.'

Much was already known about the natural world by the fourth century BCE. Aristotle's study of the weather survives in its entirety, unlike *On Nature*, written a century earlier by Anaxagoras, only fragments of which exist today in the texts of later writers. Anaxagoras grew up on a strip of coast that is part of modern-day Turkey, and moved to Athens as a young man, where he played a pivotal part in the philosophical scene. He discovered the cause of eclipses, and boldly challenged accepted ideas by claiming that the sun was not a god but a mass of red-hot metal, that the moon was earthy and that stars were fiery stones. His theories angered the citizens of Athens, and he was charged with impiety and sentenced to death. Although his life was spared, he was forced into exile in Lampsacus, a Greek city in north-western Asia Minor, where he died in 429 BCE.

As well as his theories of the universe, Anaxagoras held surprising views on the process of scientific investigation itself. After a lifetime of looking at nature very closely, he believed that our powers of perception might be deceptive and that we should always question the evidence of our eyes. He gave an astonishing example: is snow white or black? The philosopher Sextus Empiricus explains: 'Anaxagoras set the appearance that snow is white in opposition to the claim that snow is frozen water, water is black, and therefore snow is black.' If water is black, then surely snow must be black too? Cicero's report is more extreme: he writes that Anaxagoras claimed that his eyes *perceived* snow as black, as well as his mind knowing that it was. He saw black snow.

Now scientists know much more about the way our eyes perceive colour. Visible light contains the whole spectrum: red, orange, yellow, green, blue, indigo and violet. When light strikes an object, the photons may bounce back (reflection), bounce aside (scattering), pass through (transmission) or forfeit their energy (absorption). The colour of an object depends on the way light interacts with it. A holly leaf is green because it reflects the green light back to our eyes and absorbs the other colours. A lump of coal is black because it absorbs all the colours and reflects none back to the eye. Snow reflects all the light; it doesn't absorb, transmit or scatter any one wavelength more than another. So all the colours that enter a snowdrift are reflected back to our eyes, and they combine to make white light.

Snow can, however, sometimes be pink – as Aristotle discovered. In *History of Animals* he wrote: 'worms are found in long-lying snow; and snow of this description gets reddish in colour, and the grub that is engendered in it is red.' Today snow that turns red (or pink) – from algae – is known as 'watermelon snow'. And while watermelon snow takes on colour on the ground as algae grows upon it, there is now snow that is already red, orange or brown as it is falling from the sky. Any chemical in the air can change the colour of precipitation. Orange snow containing dust from storms in the Sahara fell in the Alps and in Russia during 2018, and due to the presence of dust or petroleum pollutants in the air, black snow is a reality – at least as we perceive it.

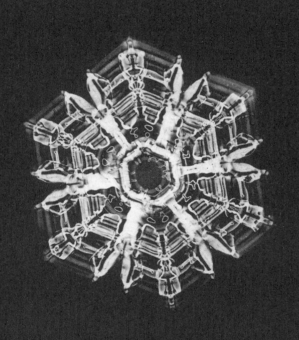

22.

Neviera

snow hut or icehouse
(Italian)

In the centuries before the discovery of electricity and the invention of the refrigerator, ingenious natural methods were developed to preserve foods and to keep drinks cold. In Mediterranean countries such as Italy, snow was gathered on the mountains and compacted into ice in a *neviera* for use in the kitchens of coastal cities. The art of cold cuisine was debated from Roman times onwards. Pliny the Elder – author of one of the first encyclopaedias – disapproved of the preservation of snow in order to chill wine. He believed snow should never be kept beyond its season, and saw it as against the natural order of things to 'turn the curse of mountains into a pleasure for the throat'.

The French philosopher and essayist Michel de Montaigne experienced the Italian fashion for chilled drinks while staying in Florence in the summer of 1581. One evening he dined at the house of Signor Silvio Piccolomini and recorded that 'it is customary here to put snow in the wine glasses. I put only a little in not being too well in body.' He was feeling under the weather because he had refreshed himself rather too much with Trebbiano wine while travelling in the heat. The use of snow and ice to cool drinks was increasing across Western Europe, with the exception of Britain.

At last the English writer and garden designer John Evelyn discovered *neviere* while travelling in Europe in the 1640s. His Continental peregrination took place long before the Grand Tour put seeing its wonders on the bucket list for eighteenth-century aristocratic youth. Evelyn's diaries and letters home are full of the novelties he experienced. He bewailed his 'melancholy and troublesome' adventures through 'an ocean of snow' while riding a mule across the Alps, where 'it snows often, so it perpetually freezes, of which I was so sensible that it flawed the very skin of my face'. By the time he reached Italy, he had had quite enough of the cold.

Evelyn wrote to his friend, Robert Boyle – one of the first experimental scientists to study cold – telling him of a curiosity he had encountered on his journey: 'snow Pits . . . sunk in the most solitary and cool'd places', often in the shade of mountains or trees. He noted that the *neviere* were built in different shapes, with the majority of the structure being beneath the earth to keep away the contents from the sun's heat. The most common design was a stone-lined cylinder underground, topped with an ornamental cupola. But the shape and materials varied: Evelyn sketched out a cross-section in his letter, showing a high pyramid-shaped thatched roof with a vertiginous pit beneath it. To preserve the snow, he told Boyle, the farmers 'beat it to a hard cake of an icy consistence, which is near one foot thick, upon this they make a layer of straw, and on that snow, beaten as before, and so continue a bed of straw and a bed of snow till the pit be full to the brim'. The straw served both as insulation and to mark out a section that could be cut during the summer.

Despite his fascination for these structures, Evelyn was cautious about consuming their contents. Like his fellow writer Montaigne, he feared the implications for his health of taking 'wine cooled with snow and ice'. 'I was so afflicted with an angina and sore throat,' he recalled after one beverage, 'that it had almost cost me my life.' However, he was seduced by the curious cold chambers, and brought his knowledge of *neviere* back to England with him. He offered icehouses to clients as part of his garden designs in London and its surrounding estates, and the fashion for 'Conservatories of Snow' soon caught on.

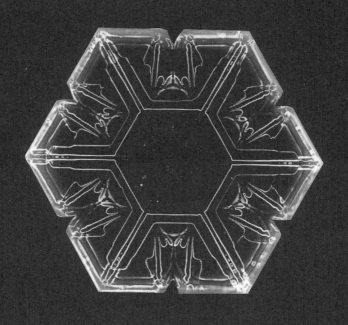

23.

Xuě qiú

snowball
(Chinese: 雪球*)

China has launched many discoveries on the world, from the materials and processes that make books possible – including papermaking around 100 CE and woodblock printing in the seventh century – to e-cigarettes in 2003. The invention of gunpowder during the ninth century and cannonballs during the twelfth century revolutionised warfare. These first cannon fired spheres of dressed stone rather than iron, and were adopted as a weapon in Europe and the Middle East during the thirteenth and

---

* The Chinese character for snow (雪) evolved visually from the ancient oracle bone script, whose image brings to mind a scene of deep snow lying over a forest. Even the modern character relates to the weather, since the top half (which looks a bit like an umbrella) is actually the radical derived from the word for 'rain', while the bottom half lends a phonetic component.

fourteenth centuries. Over time this technology led to other military developments such as landmines, hand cannon and rockets. Writing, reading and warfare never go away.

In the 1930s the world was facing an epoch-making conflict. In the Pacific and Asia, tensions between the Republic of China and the Empire of Japan caused hostilities to break out in 1937. The idealistic young Hungarian-born photographer Robert Capa, who had been covering the Spanish Civil War, viewed the Sino-Japanese War as the eastern front of a global fight against fascism. In 1938 he travelled to China, but he was kept under surveillance by the Nationalist government and never made it to the battleground. He did, however, manage to document the resistance and observe Japanese air raids on the city of Hankou, on the north bank of the Yangtze river at the mouth of the River Han. (Hankou is now merged with the city of Wuhan.) He was present for the decisive battle, which took place at Taierhchwang, a brick-walled settlement on the Grand Canal – during which Chinese troops resisted the attack of Japanese forces, in the hope of protecting the strategic Lung-Hai railway. His portrait of a boy soldier appeared on the front cover of *Life* magazine's 23 May 1938 issue under the headline 'A Defender of China'. More photographs of the conflict ran inside.

The most powerful image Capa captured in Wuhan does not show the action of war. In one photograph, children even younger than the boy soldier are exhilarated by the novelty of a light snowfall. Each long changshan makes a distinctive silhouette against the snow, which their feet have churned up, leaving a scattering of light shadows. The proud boy in the centre of the group holds up a snowball of significant size, already anticipating its target – the child in front of him bending over to collect a handful of snow to make his own. Another child looks up into the sky, a smile on his face. What would happen to these children, who even in war carried on their games? Were they aware that their playful fight was practice for a future role in the conflict? In the following year a state of total world war would emerge, directly involving more than 100 million people from more than thirty countries. It was to be the deadliest conflict in human history – with China suffering more casualties than any other nation.

No one aims a snowball at the photographer, his journalistic neutrality respected even by the children. Capa was later described by the magazine *Picture Post* as 'the greatest war photographer in the world'. But his approach to the complex subject of war was modest. 'All you could do was to help individuals caught up in war, try to raise their spirits for a moment, perhaps flirt a little, make them laugh,' he wrote. 'And you could photograph them, to let them know that somebody cared.'

24.

Snöängel

snow angel
(Swedish)

One of Stockholm's smallest but most elegant city centre parks is the steep Observatorielunden, in the Vasastan district. It's the last trace of an esker, a long mound that was formed when the last ice sheet receded from Sweden about 10,000 years ago. As the name suggests, on top of this hill is an observatory built by the Royal Swedish Academy of Sciences. The Old Stockholm Observatory is not only a site from which to look up to the stars: there's also a meteorological station that has been in operation daily since 1756, making it the oldest continuous record of temperature in the world. And for over a century, since 1904, snow depth has been measured here every winter evening on the dot of seven.

While snow is found year-round on mountain summits or at the solitude of the poles, in most Scandinavian cities snow falls only in the heart of winter – a time of short days and long nights, of storytelling and imagination. Urban snow is dependent on a particular set of timings and conditions, almost as miraculous a coincidence as falling in love. The relative warmth created by densely built-up areas, not to mention the force of traffic, can soon turn even the most generous snow flurry to slush.

Snow in the city moves time into another dimension, opens up a space for doing something out of the ordinary. There is a brooding stillness in the air before a heavy snowfall begins. The snow starts to cover up all evidence of the everyday. People eat lunch in their offices like polar explorers isolated from the world, even though the trams continue to sigh along the tracks only a few minutes' walk away. As the long afternoon wears on, snow continues to fall and the view through the window grows paler and paler and the sense of solitude increases. When the snow does lie, the urge to revel in it is powerful, and the culmination of such celebrations must be a *snöängel*.

Snöängels are written about in many Scandinavian children's books, but the impulse to make them doesn't fade with childhood. It is easy enough to find a strip of deep, inviting and undisturbed snow for a *snöängel* in the lake-strewn lowlands beyond Stockholm, but those working in the capital will need to head to the nearest park. Like an astronomer looking at the heavens, they will lie down on their backs and look up into the opaque sky. Who was the first person to spread their arms and give their snow silhouette angel wings? The shape of the human body is remade as divine just by moving our limbs, the arms waving up and down, and the legs sweeping apart to suggest the dress. Stand up, look back, and it is as if an angel fell to Earth, making an impression in the snow before walking away. Will it still be there in the morning? This time, the marks we humans leave behind will last only as long as the snow itself.

25.

Gangs

snow

(Tibetan: གངས)

To Buddhists white is the sacred colour of water, a colour symbolic of the transformation of ignorance into wisdom. Mountains capped with white snow are seen as holy, with some believed to be the homes of the gods. These are places usually sought out only by the ascetic and the mystic: discipline is required to reach the peaks, and the thin air and exposed slopes pare existence back to a minimum. Little life can survive here on the steep scree, thousands of feet above sea level.

One plant has adapted to this extreme environment, however. The delicate and dreamlike appearance of the snow lotus belies its tough survival skills. The white petals, which curl protectively around the black seed head, are enveloped in downy white hairs. The plant will flower and seed only once, and so it is all the more necessary that the dense wool protects the precious seeds from frost damage at night, and shelters them from the intense ultraviolet light of the sun during the day. Yet this plant, which has adapted so carefully to its harsh habitat, is now threatened by humans.

The snow lotus is not sought after for its beauty. Tibetans have known for centuries of its medicinal properties, and so they collect it despite its distance from human habitation. In fact, it is believed that the higher up the peaks the plants are growing, the more potent they will be. Plant collectors chant a mantra as they pick the lotus, out of respect for its sacred properties, and believing that this will increase its healing powers. They believe that the snow lotus is most potent just before it flowers and seeds. Down in the valleys, the stems are hung from the rafters to dry, before being pounded into powder to give to patients. Since there is no access to conventional medicines in the remote Tibetan villages, traditional herbal remedies are still used to treat every condition from altitude sickness to heart problems.

The healing reputation of the snow lotus has travelled to China, where it is valued as highly as the lingzhi mushroom and old ginseng. In Chinese medicine it is ascribed a nature both bitter and sweet, and classified as a herb that invigorates the blood; it is used to treat back pain, rheumatism, and ailments associated with pregnancy and menstruation. The dried blossoms are in demand by apothecaries to target the spleen, the kidneys and the liver. Whereas practitioners of traditional medicine brought down as many flower heads from the mountain slopes as were needed, now the high regard in which the snow lotus is held is a threat to its survival. People who are plant-hunting for profit do not set any limit on how much they collect, and they take the biggest blossoms (which are easier to find and fetch more money on the market). In consequence, the smallest plants are the only ones left to reproduce, and this has caused the average height of one species to diminish by several inches during the last century. While the aquatic lotus, which rises from deep lakes, has long been revered in Buddhism as a symbol of purity and rebirth, the snow lotus of the high mountains tells another, more cautionary story: scarce in scope and shrinking in size, it counsels humans to learn to take only what they need from nature, so that no one will lack its benefits in the seasons to come.

26.

Calóg shneachta

snowflake
(Irish)

Ireland gave the world one of the most famous mentions of snow in literature. 'The Dead', the final short story in James Joyce's collection *Dubliners*, records party small talk about the weather:

'They say,' said Mary Jane, 'we haven't had snow like it for thirty years; and I read this morning in the newspapers that the snow is general all over Ireland.'

'I love the look of snow,' said Aunt Julia sadly.

'So do I,' said Miss O'Callaghan. 'I think Christmas is never really Christmas unless we have the snow on the ground.'

Although the story is written in English, the question of the Irish language runs through it. One of the guests, Gretta Conroy, grew up in the west of Ireland, the Gaeltacht, where Irish was spoken even after the Anglo-Norman invasion. As the party ends, a song recalls to Gretta a young man – Michael Furey – who once courted her in Galway. She suspects he died for love – because he stood a whole night outside her window in the snow.

Now, as the snow falls, Gretta's husband Gabriel muses on her grief, and considers the influence the dead exert on the living.

Yes, the newspapers were right: snow was general all over Ireland. It was falling softly upon the Bog of Allen and, further westwards, softly falling into the dark mutinous Shannon waves. It was falling too upon every part of the lonely churchyard where Michael Furey lay buried. It lay thickly drifted on the crooked crosses and headstones, on the spears of the little gate, on the barren thorns. His soul swooned slowly as he heard the snow falling faintly through the universe and faintly falling, like the descent of their last end, upon all the living and the dead.

Softly, thickly drifted, falling faintly; would those soul-swooning snowflakes have drifted any differently had Joyce written in Michael Furey's Irish? *Dubliners* was published in 1914 after Joyce had left Ireland to live in Europe. Irish was then an unknown language for most living in the capital city. Gabriel is taken to task by a dance partner for the lack of political engagement in his newspaper column, which reviews Browning's poetry and other Anglophone works. In Joyce's next book, *A Portrait of the Artist as a Young Man*, Stephen (his fictional alter ego) considers the issue during a conversation with the dean at his college. 'The language in which we are speaking is his before it is mine. How different are the words *home, Christ, ale, master*, on his lips and on mine! I cannot speak or write these words without unrest of spirit. His language, so familiar and so foreign, will always be for me an acquired speech. I have not made or accepted its words. My voice holds them at bay. My soul frets in the shadow of his language.'

Snow, too, sounds different in Irish. Snowdrifts (*sneachta*) are forged of individual snowflakes (in the plural, *calóga sneachta*). *Sneachta* is related to other European snow words: *Schnee, sneeuw, sněžení, sne, snow, neige, nieve*. Less commonly used compound terms, many of which include words relating to textiles, offer more specific descriptions: *bratóg* (rag) *shneachta*; *cáithnín* (a small particle) *sneachta*; *fíneog* (a mite) *shneachta*; *lubhóg* (flake or drop) *shneachta*; *slám* (a tuft or handful) *sneachta*; *spitheog* (a smidgeon) *shneachta*.

Joyce's vision of a nationwide white-out is built up from a trillion individual snowflakes – just like the words that the writer assembled in his epic manuscripts, or indeed as learning any language requires a slow accumulation of vocabulary. Since Ireland gained independence from Britain in 1922, its government has been striving to re-introduce Irish; the nation is now bilingual, and Irish is an official language of the European Union.

27.

Huka-rere

snow, one of the children of rain and wind
(Māori)

During the last millennia, Polynesian explorers crossed thousands of miles of turbulent Pacific waves in beautifully carved wooden *waka* or outrigger canoes. They grew to know a network of small volcanic islands that lay many days of travel apart. In the hull of the vessel lay a 'wave pilot', a person who navigated using their knowledge of the pulse of ocean currents beneath them and the movement of winds and stars above. The pilot held the pattern of the seas in their memory, for the charts were irreplaceable artefacts that were usually left behind on land. In these charts, the ocean swells were represented by slim sticks, lashed together at the meeting point of tides, and tiny white cowrie shells symbolised the islands. These were difficult, dangerous journeys, even with an understanding of wayfaring. Keen eyes were also needed to spot land on the horizon, or to predict changing weather conditions. Clouds, in particular, were a means to tell a forecast. The travellers looked out for *titi taranaki* or radiating streaks of cloud, a sign of bad weather; a bank of clouds, *he whare hau* ('a house of wind'), which indicated blasts were on their way; clouds lying in layers above the horizon, *rangi mātāhauariki*, the forerunner of the cold south wind; and a sheet of mist evenly covering the sky, *papanui*, a signal that it would be calm the next day. They hoped not to see *kaiwaka*, threatening billows on the horizon, and a sign of misfortune. Perhaps the most beautiful sight when voyaging at night was *pīpipi* – cirrostratus clouds. These high curtains of ice crystals are transparent, so that the full moon can be easily observed through them – indeed, sometimes the only clue to their presence is a halo around a bright planet.

Clouds do not only herald unwelcome storms and winds: the first sign of land from a boat is often a fixed cumulus cloud in an otherwise clear sky. Finally, these exploratory island-hopping voyages in the South Pacific led the seafarers to a new shoreline in the far south, characterised by dramatic mountain ranges. (The date East Polynesians arrived in New Zealand is a matter of debate, but is usually given as the thirteenth century.) They named this land Aotearoa, which means 'long white cloud'. It is believed that this was coined because of the distinctive standing waves of lenticular clouds (*ao*) that rise from the mountains.

Another explanation is that from a distance the travellers might have mistaken unfamiliar snow on the peaks for the clouds they were more accustomed to reading.

The tribes brought with them to Aotearoa their own myths about the creation of the world and the meteorological forces that so influenced their lives. They knew everything in nature was interconnected: the whole great pantheon of Māori gods is descended from Ranginui, the Sky father, and Papatūānuku, the Earth mother. One of their sons, Tāwhirimātea, governed the weather, creating thunder and lightning, wind clouds and storms. (To convey the close relationships between the elements, the new inhabitants of Aotearoa had another kind of chart, just as ingenious as those they created to navigate the seas. They drew up complex genealogies, or *whakapapa*, which outline a great planetary pantheon in which divine offspring personify distinct weather features.) Since the weather in this part of the world is mercurial, and characterised by frequent rain and strong winds, a rich store of words, sayings and trad-itions relate to the domain of Tāwhirimātea. One of the most significant tales relates his vengeance on his sixty-nine brothers who had separated his parents – the Earth and the Sky – from an eternal embrace, wrenching them apart to create more space for themselves. In his fury, Tāwhirimātea turned on his brothers and the aspects of the world they represented – he destroyed the for-ests, drove the sea god and his offspring into the sea, and wreaked havoc on crops. Tāwhirimātea's own children, who were winds and clouds of different kinds, all participated in these battles – Apū-hau, Apū-matangi, Ao-nui, Ao-roa, Ao-pōuri, Ao-pōtango, Ao-whētuma, Ao-whekere, Ao-kāhiwahiwa, Ao-kānapanapa, Ao-pākinakina, Ao-pakarea, and Ao-tākawe (the offspring represented, in turn, 'fierce squalls, whirlwinds, dense clouds, massy clouds, dark clouds, gloomy thick clouds, fiery clouds, clouds that preceded hurricanes, clouds of fiery black, clouds reflecting glowing red light, clouds wildly drifting from all quarters and wildly bursting, clouds of thunderstorms, and clouds hurriedly flying on').

There may be an abundance of clouds over Aotearoa, but only a few bring snow – which gets a brief mention in a myth told by the Takitimu tribe. The spirit of rain, Te Iho-rangi, was responsible for controlling the clouds, and ensuring that they always provided a protective veil between Ranginui, the Sky father, and Papatūānuku, the Earth mother, shading the body of the Earth. The Takitimu believe that Te Iho-rangi and the spirit of wind, Huru-te-arangi, became lovers. Their offspring, only twelve in number, were known as the Snow Children and the Frost Children: Huka-puhi, Huka-rere, Huka-papa, Huka-taraapunga, Huka-waitara, Huka-waitao and Huka-puwhenua, Huka-punehunehu, Huka-pawhati, Huka-rangaranga, Huka-koropuku and Huka-teremoana. The word *huka* is still the core of a number of compound words in the Māori language (*te reo Māori*) for cold conditions such as *huka-rere* (snow), *huka-papa* (ice, frost) and *huka-waitara* (hail). Meanwhile, these 'Children of Cold' continue their chill existence in the realm of Paraweranui, the fierce and stormy south wind.

28.

snowboarding

(American Sign Language)

Snow is often associated with silence. When a deep layer of fresh, light snow lies on the ground, sound waves are readily absorbed at the snow surface, dampening the world's din. But over time, and in changing weather conditions, the quality of the snow alters. For example, if the surface melts and then refreezes, becoming smooth and hard, it will reflect sound waves. Noises made nearby may be enhanced and travel farther. Silence is relative.

A layer of snow is made up of many tiny crystals surrounded by air pockets, which are compressed when someone steps or slides over them. As they squeeze together the ice grains rub against each other. This creates friction and the lower the temperature, the greater the friction between the grains, which produces the familiar creaking or crunching sound. Nobody knows at what temperature snow starts to crunch, but the colder the snow, the louder the crunch. Thus, when a snowboarder makes their swift descent down a pristine mountain slope, they create a soundscape as well as marking the run with their tracks. In between jumps, grabs and other aerials, the way the snowboarder moves on the surface of the snow, managing rides and slides, carving and tapping, creates a satisfying shredding or scrunching sound – even a roar.

For some champion snowboarders, sound is irrelevant when it comes to pushing the human body and mind to its limits. Take Lauren Weibert, a stellar athlete who claimed a gold medal in the women's slopestyle event and a silver in the snowboard cross event at the 2015 Deaflympics held at Khanty-Mansiysk in Russia, who again won gold in the women's slopestyle at the 2019 Deaflympics in Valtellina-Valchiavenna, Italy. Yet it took snowboarding a while to join the ranks of the other winter sports at the games – only making its first appearance in 1999. Since then, US snowboarders have won nearly thirty medals in snowboard events alone. The US first entered skiing at the Deaflympics in 1967, and came home with two gold medals. The first national meeting of Deaf skiers was held in 1968, a time of wider social change. The US Deaf Skiers Association was formed soon after, but it was not until 1998 that the organisation's name was changed to include snowboarders – it is now known as the US Deaf Ski and Snowboard Association (USDSSA).

Anyone who wants to learn how to sign 'snow' in American Sign Language will easily find video instruction online, the sign described as: 'Both open hands move down in slow wavy movements while fingers wiggle.' There is more debate around the sign for snowboarding, which is a relatively recent addition to the language; the development of the sign reflects a growing understanding of the sport. According to experts at American Sign Language University, the original version of 'snowboarding' was the sign for 'snow' followed by the sign for 'surfing'. This parallels the evolution of snowboarding terminology in English, which often derives from the language of skaters and surfboarders. Later, the sign for 'ski' was conscripted, in which 'flat hands' slide forward repeatedly. The most recent version (depicted above, with two quick movements forward and down, at an angle, as if sliding down a mountain) truly reflects the debonair dynamic of a snowboarder on the slopes.

29.

Kava

snow, snowfall
(Faroese)

In the middle of the North Atlantic between the Shetland Isles and Iceland, the warm currents of the Gulf Stream meet the cold waters of the Arctic. Here, at the heart of the global climate and weather machine, lie the wave-lashed Faroe Islands. The location means that this nation of eighteen major islands – plus many islets and skerries – has a climate that shifts like quicksilver. In one day it's possible to experience weather associated with each of the four seasons. Winters are mild, and snow (*kava*) is not as common as might be expected at these latitudes. Even so, the Faroese–English dictionary contains a wealth of related terms: on a *kavadagur* (day with snow) it falls from a *kavaluft* (snow-filled sky) onto a *kavabrekka* (snow-covered slope). Snow cover is generally thin, even on the exposed ridges – a gauze veil through which the rich tones of the underlying scree can still be seen.

The tides that sweep around the archipelago have brought many travellers, and the Faroese language has evolved over the centuries to reflect the influence of settlers from different shores. In the Viking age, Old Norse arrived with the Norsemen, and Gaelic came with the *papar* or monks. Faroese words will often be familiar to speakers of other Nordic languages, but Faroese has a distinctive character. It has proved to be as resilient as the people who live here. After the Reformation in 1536, the ruling Danes outlawed the use of Faroese for education, religious services and administration, and for almost three hundred years Faroese was not written down. However, the islands' residents continued to use the language every day in conversation, songs and folktales.

It was a Lutheran minister called Venceslaus Ulricus Hammershaimb (1819–1909) who at last developed a spelling system for Faroese in 1846. The drive to celebrate the modern language in print and on paper marked a new phase in the nation's identity: the movement for independence from Denmark. A new national flag was also needed. The *Merkið* ('banner') was dreamed up in 1919 by three students from the Faroes living abroad in Copenhagen, Jens Oliver Lisberg, Janus Øssurson and Thomas Pauli Dahl, and the first sample was sewn by Ninna Jacobsen. As with the language, the flag is similar to those of neighbouring Scandinavian nations. The other 'Nordic Cross' flags contain the colours white, red and blue in different layouts – and since the Norwegian flag has a red background while the Icelandic has blue, it fell to the Faroese cross to be laid on a white colour. The designers of the flags of some other Arctic and Baltic nations, such as Estonia and Greenland, purposefully chose white to reflect the significance of snow and ice in the landscape. (The Estonian flag can be seen to symbolise a blue sky above a frozen lake, the Greenlandic flag a red sun rising over a glacier.) The white in *Merkið* was purely a matter of chance.

Although the flag might not have been daring in its design, hoisting it was a radical act against the Danish authorities. Fishermen who sailed under *Merkið* were fined and in some cases banned from the seas. In 1948, a century after Faroese was first written down by the minister, the Home Rule Act came into force, giving the Faroe Islands the power to self-govern. Faroese was recognised as the islands' official language, and today it is spoken by over 72,000 people. And *Merkið* can be seen flying in front of many homes, even in winter, a white glint against the stormy skies like a seabird's wings.

30.

Kardelen

snowdrop
(Turkish)

A mist of snowdrops is a welcome sight at the end of winter across Europe and the Middle East – from the Spanish and French Pyrenees in the west to Iran in the east and south to Turkey and Syria. The flower is known as a 'drop' of snow in English as well as in many other languages, a term believed to be derived from the German *Schneetropfen*, the teardrop-shaped pearl earrings popular in the sixteenth and seventeenth centuries. (Other common English names such as 'Candlemas bells' and 'Fair maids of February' refer to the season when the flowers appear.) Yet not every culture names the plant for its pendant flowers; for some it is the emerging shoots that determine the name, as in French (where the plant is known as *pierce-neige*) and Turkish, in which the name is a compound of *kar* ('snow') and *delen* ('which broaches').

At least eleven of the world's twenty or so snowdrop species are found in Turkey. Their sharp, bright leaves and white flowers, the inner petals often marked with curious viridian code, can be spotted at the edges of shady woodlands, near streams and on mountain slopes. In 2019 a new species of snowdrop (*Galanthus bursanus*) was discovered growing in red clay soil in a forest near to the city of Bursa. Here, to the south of the Sea of Marmara and on the Black Sea coast, snow falls in winter but rarely lies long. The new species is unusual – it blooms in autumn, ahead of the snows, and the flowers have a delicious fragrance. The conservation status of the new species – of which only a few clumps were found – was immediately assessed as Critically Endangered.

This new species is not the only *Galanthus* under threat. The snowdrop is the most heavily traded of all ornamental bulbs collected in the wild. The bulb trade from Turkey, and in particular the Taurus mountain range, dates back to the 1550s when plants such as tulips and fritillaries began to be popular among gardeners in Europe. The commercial trade in exported bulbs increased, until by 1984 over 80 million bulbs were being exported annually. There was concern. From 1985, wild-flower bulb collection in Turkey was banned for five years. Yet today millions of snowdrop bulbs are legally exported each year, heading for domestic gardens overseas. According to the language of flowers, the snowdrop is a symbol of hope and consolation, since the green shoots are the first to pierce the snow in late winter, and the flowers signal new life every spring. In their genesis beneath the soil, these plants connect our gardens to other places – as long as indulging the brief flourish of our own hopes does not lead to the detriment of future flowers elsewhere.

31.

Omuzira

snow

(Luganda)

At the beginning of the eighteenth century, with the discovery of the 'sublime' as a new type of aesthetic experience, a European appetite grew for accounts of mountaineering expeditions, filled with descriptions of the tranquillity and terror the peaks inspired. Genteel readers imagined themselves into the mountaineers' uncomfortable hobnailed boots and sweaty cable-knit jumpers through illustrations to books of exploration in the form of engravings. Soon they yearned for more ways to share the experience of standing on these remote summits, which few would ever realise in real life. Fortunately for armchair travellers, the birth of mountaineering coincided with the development of photography.

The first summit photographs were taken in July 1861 by the French photographer Auguste-Rosalie Bisson (1826–1900), who exposed three negatives on the slopes of Mont Blanc, having climbed over 4,000 metres above sea level. The enterprise had required meticulous planning. Before the development of roll film and hand-held cameras in the 1880s, images were captured on wet glass plates that were chemically coated and developed immediately in a light-proof tent. In addition to the usual challenges the peaks presented to climbers, it was a logistical problem to manoeuvre the heavy photographic equipment up the slopes. Extra porters were conscripted to carry the plates, cameras, chemicals and canvas 'darkroom'.

Some of the most striking images from this era were taken by legendary photographer Vittorio Sella (1859–1943) in eastern equatorial Africa. Sella's father had written the first Italian language treatise on photography in 1856 and his uncle Quintino Sella, a distinguished statesman and keen Alpinist, founded the Italian Alpine Club in 1863. Sella accompanied Luigi Amedeo, the Duke of Abruzzi, on his 1906 expedition to the Rwenzori mountain range. The duke was making a name for himself as an explorer: in 1897 at the age of twenty-four he had organised and led the expedition that made the first ascent of Mount St Elias in Alaska, and as the century turned he had set off in a converted whaling vessel named *Stella Polare*, in an attempt to reach the North Pole. Unfortunately, severe frostbite in his fingers obliged him to abandon the command of the expedition to others.

Now the duke set his sights on the 120-km-long Rwenzori range on the border between Uganda (at the time a British protectorate) and the Democratic Republic of the Congo. Rwenzori means 'rainmaker' and the melting snow and glaciers of Rwenzori provide a small amount of the River Nile's water. The expedition (the core team of which included a geographer, a geologist, a botanist and a cook) travelled from Naples to Mombasa by boat in late spring, and then marched from Lake Victoria, so far from its objective that the Rwenzori were not visible for several days. It was to be a journey of months, not weeks, which required significant supplies of food and clothing, and the duke's entourage is reported to have stretched along the road for half a kilometre. The enterprise was only made possible by the work of many Baganda porters, who would have spoken Luganda, a Bantu language and one of the main languages of Uganda. History does not record whether the duke learned how to say *mwattu* ('please') and *oli otya* ('how are you?') to the porters depicted in Sella's photographs.

From the hills north of Kaibo, which make a watershed between the lakes Albert and Edward, everyone walking in the caravan could see the distant mountain range for which they were destined, emerging from the mist and tipped with *omuzira*. At last on 29 May, the expedition reached Fort Portal, the last British outpost before the mountains. Here they exchanged their light clothes for jackets and trousers more suitable for climbing. On 1 June the expedition began a seven-day walk towards the slopes of Rwenzori, headed for Bujungolo, their future base camp. Slowly the vegetation and countryside began to change, as if they were walking from spring into summer. Here and there the first lobelias appeared, and lush dracaena palms, yet far above them the duke and his companions could see glaciers sparkling. Passing through the Mobutu Valley, they noted: 'Trunks and branches entirely covered with a thick layer of mosses which hang down in long beards from all the branches; they enlarge and fill out the knots in the wood making the plants appear strangely distorted; . . . There are no leaves except on the highest branches, but the forest is dark due to the dense intertwining of trunks and boughs.' Taking fewer and fewer men at each stage of the journey, the duke ascended slowly towards the snowline and its 'rocks entirely covered with glittering crystals, in the shape of vitreous efflorescences'.

The reader of accounts of expeditions such as the Duke of Abruzzi's grows used to a routine of early starts and remote objectives. Mountaineers needed an understanding of how to negotiate the snow, whether compact snow, which could be cut into steps with strong blows from an ice axe, or wide crevasses bridged by no more than a ribbon of ice. Sella travelled in these perilous conditions with unwieldy photographic plates measuring 30 × 40 cm, his preferred medium, although less bulky and more resilient technology would have been available at the time. He had invented equipment, including modified pack saddles, to allow the large glass plates to be transported safely on horseback. Heavy kit was not the only challenge he had to overcome. He worked in paralysing cold, subject to the nausea and vertigo of high-altitude exploration, while being constantly alert to avalanches and other dangers. Even then, the weather did not always perform for the camera. Roberto Mantovani notes:

> On June 11th while the Duke's team returns to Bujungolo camp, Sella and his companions climb first to the pass and then to King Edward Peak with the idea of taking some beautiful landscape photographs. The weather is terrible but finally, in the early afternoon, it clears. Next morning Sella, Botta and Brocherel are again on the pass and climb to Moore Peak (4654 m); an easy climb but made treacherous by the ice. It is snowing on the peak and unfortunately the camera cannot be used. On returning to the pass the roped party meet Dr Roccati and a guide. By now it is snowing heavily everywhere, and only Botta and Sella have the courage to pass the night in the tent, insisting on remaining at high altitude in the hope of good weather.

The next day was cloudy, and the vague outlines of the highest peaks appeared through the fog above Sella for only a few moments. The north face showed 'a peak and ridges edged with the most majestic cornice imaginable, sustained by countless stalactites and ice needles which, from a distance, seemed a gala of snow-white lace'.

There is a self-conscious theatricality to some of Sella's photographs. The figures of explorers are silhouetted against the snow-clad slopes in carefully orchestrated poses: they are there to give a helpful impression of scale, as well as staking a human claim to the mountain. Sometimes Sella's companions would repeat their climbs in the most dramatic locations, in order to stage a scene for the camera once the route had already been successfully accomplished. These photographs bore witness to an ascent for geopolitical ends, in a time when mountaineers were accompanied by a national flag rather than a hashtag. On reaching the highest peak, the Duke of Abruzzi unfurled the small Italian flag given to him by Queen Margherita before his departure.

Yet in Sella's other photographs, which have no humans in the frame, his fascination for the beauty of the Rwenzori is apparent. One of his iconic images is a view of the peak of Mount Stanley, the third highest mountain in Africa at 5,091 metres, on a cloudy day. Snow seems to fade to the same tone as the sky, which hangs low, threatening more snow. Snow clings to all but the most precipitous rock faces and undulates over uneven outcrops on the summit. Once intended to celebrate human achievement and magnify the glories of the natural world, today Sella's images serve another function – a means for scientists to make comparison of snow cover on the Rwenzori over time. For the snow depicted in those photographs from a century ago has long gone. In 1906, the Rwenzori had forty-three named glaciers distributed over six mountains with a total area of 7.5 km² – half the total glacier area in Africa. By 2005, less than half of these were still present, on only three mountains, with an area of about 1.5 km². A team of scientists, working on modelling climate change, have compared Sella's photographs with their own data, to predict that by 2030 there will be no glaciers left on the 'rainmaker'.

32.

Fokksnø

wind-transported snow
(Norwegian)

The noun *ski* has its origins in the Old Norse word *skíð* meaning 'cleft wood' or 'stick of wood'. Prehistoric skis preserved in bogs and ancient rock paintings show that hunters and trappers used skis at least 5,000 years ago. Earlier still, as the glaciers retreated, Stone Age hunters followed herds of reindeer and elk northwards from central Asia, travelling on skis covered with fur. Skis were then adopted across the Eurasian Arctic, and as the medieval era dawned, many thousands of years later, Scandinavian farmers, hunters and warriors were transporting themselves in the very same way.

Skiing plays a crucial role in an episode of the Saga of Håkon Håkonsson, which took place during the tempestuous civil war era in Norway. After the death of King Håkon Sverresson the king's followers were on a mission to rescue his baby son and sole heir, who had been born in enemy territory. On a cold night in January 1206, the child was snatched by a group of loyal soldiers well used to the perils of the wilderness. They were known as *birkebeiners*, or 'birch legs', because they were so poor their shoes and gaiters were made of birch bark. A speedy escape was essential, but the party encountered a strong blizzard as they approached the mountains. Storm winds can break up snow crystals and carry them some distance as they are falling. The resulting snow is known in Norwegian as *fokksnø* – although now it is often referred to in the direct translation from English as *vind-transportert snø. Fokksnø* is deceptively smooth to ski on but it can be perilous, since it does not bind with the older layers of snow beneath it on mountain slopes. At the slightest disturbance it can rebel from the snowpack, causing an avalanche.

Undaunted, two of the best *birkebeiner* skiers, Torstein Skevla and Skjervald Skrukka, took prince Håkon and sped through the snow over the mountain from Lillehammer to Østerdalen. They made it to safety in Nidaros (now Trondheim), the first capital city of Norway's Christian kings. Following his dramatic rescue, Håkon IV went on to rule for forty-six years, and founded a dynasty. The escapade that saved his life has inspired the modern-day Birkebeinerrennet ski marathon, which retraces the journey taken by Skevla and Skrukka from Rena to Lillehammer. During the 54-km route skiers carry a weight on their back to symbolise the baby prince.

In the Oslo Ski Museum there hangs a romantic oil painting

by Knud Bergslien that depicts the pair of bearded *birkebeiners* clad in elegant chainmail – an unlikely outfit, considering what is known of their poverty. The artist has paid particular attention to the detail on their skis, right down to the ornate curved tips. Of course, the shape of a ski influences a journey just as much as the quality of the snow. In the centuries since Håkon's reign, skiing has continued to be central to the Norwegian lifestyle, and Norwegians are among the world's best ski designers. In the early nineteenth century, a thin, cambered ski was developed by woodcarvers in the province of Telemark, so that the skier would not sink into the snow. In 1868, Sondre Norheim demonstrated a sidecut that narrowed the ski underfoot while keeping the tip and tail wide; like the camber, the sidecut produced a flexible ski, with an edge that could follow the shape of a turn instead of skidding sideways. Then in 1882 the very material of which skis were made changed, from ash to hickory wood. (Hickory had once been too hard to work with traditional hand tools, but new carbon-steel tools were stronger.) The dense wood made a thinner, more durable yet flexible ski, the base of which was less likely to wear and slow the skier down. Hickory wood was imported to Norway at some expense from Louisiana, until Norwegian immigrants in Wisconsin and Minnesota realised that with easier access to lumber stocks, they could make the skis – for export – far more economically in the United States. And so, just as in the distant past, skis became emblematic of the story of human movement and migration.

33.

Sniegas

snow

(Lithuanian)

Although snow does not usually fall until January now, Lithuania once lay under deep drifts from November to March. Snow cover was vital in protecting cereal crops such as winter rye from night frosts; if there was no snow, the crops would freeze. Hence the saying: *naudos, kaip iš pernykščio sniego* – 'as much use as last year's snow' (i.e. once snow has melted, it is of no use to anyone). The weather of this Baltic state has played a significant part in European history, proving useful to some armies, and ruinous to others. During the Crusades, the great marshes and wide lakes became frozen battlegrounds, on which knights in their heavy armour were in as much danger from thin ice as from enemy weapons. And several centuries later, the capital, Vilnius, saw the bitter denouement of Napoleon Bonaparte's Russian campaign.

The French emperor had entered Moscow in September 1812 believing himself a conqueror – even though his Grande Armée, which originally comprised more than a million men, had been decimated by starvation and bloodshed on the long march across Europe. He found the great city almost deserted. The day after Napoleon's arrival fires were started and many buildings, including those earmarked for his troops' winter quarters, were burnt to the ground. Nevertheless, he stayed in Moscow waiting for Emperor Alexander I – who was in St Petersburg – to make him an offer of peace, which never came. Late in the year, he realised it was implausible to remain longer in the ashes of the ruined city, and ordered his army back on the road to France.

On this retreat through November snow, the exhausted and poorly clad troops were attacked by Cossacks. Their horses ran out of fodder, and struggled to walk on the icy roads since the army lacked the equipment to forge shoes for them. The Marquis de Caulaincourt, who accompanied Napoleon as Grand Écuyer or Master of the Horse, had advised against the campaign. He recorded his horror, on seeing men fall by the road unable to rise again:

Once these poor wretches fell asleep they were dead. If they resisted the craving for sleep, another passer-by would help them along a little farther, thus prolonging their agony for a short while, but not saving them, for in this condition the drowsiness engendered by cold is irresistibly strong . . .

I tried in vain to save a number of these unfortunates. The
only words they uttered were to beg me . . . to go away and
let them sleep.

After a journey of almost two months, as few as 50,000 stunned
survivors of the Grande Armée struggled across the icy Berezina
river using improvised bridges, and dragged themselves in freezing winds towards Vilnius.

The officer François Dumonceau described the scene outside
Vilnius's city gate: 'a veritable moving mountain, more than two
metres deep, of dead and dying, pushing, shoving, hemmed in
on all sides, at each step risking being thrown down by the con-
vulsive spasms of those we were trampling underfoot'. As many
as half of the starving survivors who managed to reach Vilnius
died. Some succumbed to exhaustion or cold on arrival; others
may have over-eaten, in their desperation to assuage their raven-
ous hunger, or drunk themselves into oblivion. Some soldiers
had frostbitten noses, toes or fingers, which turned gangrenous;
the wounded soon spread sickness to the townspeople. As the
lodging-houses were full, the unluckiest froze to death in the
snowy streets. More than 400,000 men lost their lives during
the winter invasion. The campaign was a turning point in the
Napoleonic Wars, and a significant blow to Napoleon's ambitions
of European dominance. Within sixteen months, the emperor was
exiled in the balmier climate of the island of Elba.

34.

Sira

snow

(Tundra Nenets: сыра)

The reindeer of Siberia migrate more than a thousand kilometres every year in search of lichen pastures, and the Nenets
people accompany them. They travel in all seasons, starting in
the south-west of the Yamal Peninsula as the spring sunshine
breaks through the forest tundra, and reaching the coast of the
Kara Sea late in the summer. In November, as nights draw in and
snow begins to fall, the reindeer retrace their tracks southwards.
The winters in this region are severe, and the route is difficult – it
includes an epic 48-km crossing of the frozen Ob river – but the
reindeer are strong and accustomed to challenging conditions. In
fact, reindeer can cover the great distances faster on the winter
journey, travelling between 8 km and 20 km a day when there is

snow (сыра) underfoot, and between 3 km and 11 km in summer when the ground is softer, and sometimes swampy.

Snow dictates the Nenets' lifestyle over the return journey, and they depend on reindeer for food, clothing, tools and even shelter. They wear warm furs to keep out the cold, and live in pyramid-shaped tents, or *chums*, made of pale reindeer hides, which can be struck and packed up on wooden sleds whenever the camp needs to move on. The reindeer work as draft animals, pulling the sleds that carry the *chums*, huge bundles of firewood and sometimes even transporting people. One sled is set aside for sacred objects such as votive bear skins, coins and the small wooden figures representing each family's ancestors. The herders' religion is animist and shamanistic, and respect for the land and its resources is never left behind.

'Nenets' is used of a pair of related languages – Tundra Nenets and Forest Nenets – but the two are very different and speakers cannot understand each other. Most Nenets people will also speak Russian, since from the late Stalin era onwards they were placed in Soviet boarding schools where Russian was the primary language; UNESCO classifies Nenets as an endangered language. A gradual shift in language use is not the only influence the wider world is having on traditional Nenets ways of life. The reindeer herds are in decline. The tundra landscape is changing fast. The climate is warming as a result of human fossil-fuel consumption, and Russia is on a quest for further reserves of hydrocarbons.

Sometimes extraction takes place in the midst of traditional migratory routes: herders must lead their animals under pipelines or across roads, as on the Bovanenkovo gas field. Gazprom, which operates Bovanenkovo, lays down a geotextile known as the 'white carpet' to protect the sledges from the tarmac.

The damage to the environment underground is less easily resolved. As the permafrost begins to thaw the landscape is metamorphosing in eerie ways: sinkholes suddenly appear where gas has exploded underground; in other places, the surface of the earth appears to vibrate, a phenomenon likely caused by bubbles of methane and carbon dioxide bursting to the surface. Petrochemicals spill into rivers, and flow towards the Arctic Ocean, as in the major incident in the Pyasina river in June 2020. The region is suffering hot summers without rain, making the ground dry and difficult for the reindeer to draw the heavy sleighs across. In winter, the snow is no longer dependable: heavy rainfall on top of snow is becoming more frequent, and when this ices over, is it almost impossible for the herds to graze. The Nenets' characteristic resilience is being tested in new and rapidly changing ways.

35.

Taccuqt

snow

(Tamazight: ⵜⴰⵛⵛⵓⵇⵜ)

The story of the city of Ifrane in Morocco's Atlas Mountains is one of contrasts: deep snow in the desert hills; intense cold on the Mediterranean coast. *Yfran* means 'caves' in Tamazight, one of the main languages of North Africa before the Muslim conquest of the seventh century brought Arabic to the region. The Atlas Mountains have long been home to the Amazigh, indigenous people of North Africa (commonly referred to as Berber), whose traditional territories stretch from the Mediterranean in the north to the Sahara in the south. During the cold months in the Atlas, caves were used by Amazigh herders as temporary dwellings once they had packed up their tents and brought their flocks of goats and sheep down from the peaks to lowland pastures away from the snow or *taccuqt*. When the first year-round settlement in the Tizguit Valley grew up during the sixteenth century, the community lived in caves hollowed out of the limestone valley wall. The village was called Zaouiat Sidi Abdeslam.

Several centuries later, on land a few miles upstream, the French colonial administration established the modern town of Ifrane. This 'hill station' was designed for European diplomats and their staff to find relief from the scorching heat of cities like Fez and Marrakesh in summer. The year was 1929, and the 'garden city' model was in vogue in Europe. Befitting Ifrane's location, on snowy peaks more than 1,700 metres above sea level, the architect's plan called for chalet-type summer homes in the Alpine style: 'Maison Basque', 'Jura' and 'Savoy'. These were set among gardens and winding streets lined with lilacs and lindens and chestnuts, named after the trees themselves: Rue des lilas, Rue des tilleuls. The botanical theme was continued in the naming of iconic establishments like the Perce Neige Hotel ('snowdrop hotel'). When Ifrane was still a very new city, it achieved notoriety by breaking the record of the lowest temperature ever observed in Africa: minus 23.9° Celsius on 11 February 1935. But the picturesque facades of Ifrane masked an oppressive colonial reality – the garden-city plan allowed no space for the population of local people employed in the service of the French as maids, gardeners, cooks and more. A shanty town called Timdiqin grew up on the other side of a deep ravine, to house the workers – with a very different kind of architecture.

After independence in 1956 the fine French properties in the original garden city were gradually bought up by Moroccan families. Ifrane was granted a mosque, a market and new housing estates and the neighbourhood of Timdiqîn was rebuilt with proper civic amenities. Now Ifrane is a resort town, a 'little Switzerland' that attracts tourists from Morocco and beyond. There's even a yearly snow festival, which welcomes competitors for the crown of 'Miss Snow of Ifrane'. But inequality lives on beyond the city limits: in recent winters the snows that entertain the wealthy pleasure-seekers of Ifrane have cut off the Amazigh living in underdeveloped rural areas in the High Atlas from vital supplies, causing frequent fatalities from cold, hunger and a lack of medicine.

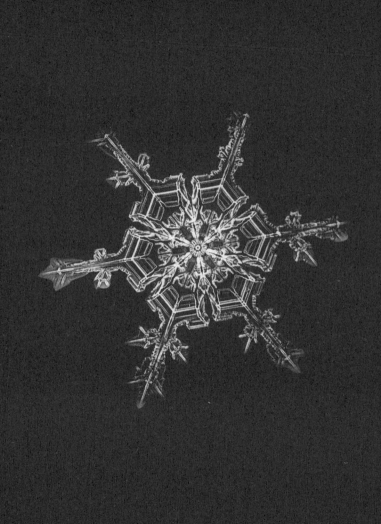

36.

Himá

snow

(Sanskrit: हिम)

On images of the Earth captured by satellites from space, the Himalayas look like delicate fern fronds or frost flowers, unfurling from jungles north of Burma to arc along the borders of India and Tibet, nodding over Bhutan, Sikkim and Nepal, and reaching up to the dusty glaciers of the Karakoram on the frontier between Pakistan and China. These silver fronds are the enormous tongues of 15,000 glaciers, including the great Siachen Glacier, the largest outside the polar regions. All 1,500 miles of the Himalayas are snowbound throughout the year, despite being so close to the tropics, and the snowline – at around 5,500 metres – is one of the highest in the world. In fact, the name of the range derives from the Sanskrit word for snow, *himá* (हिम), with *ā-laya* (आलय, 'receptacle, dwelling'), thus *Himālaya* (हिमालय): 'abode of the snow'. In Hinduism the mountains find personification as the 'king of snow', Himavata (हिमवत्, literally *frosty*), ruler of the Himalaya Kingdom of Ancient India and father of Parvati, goddess of fertility, love and beauty.

Millions of years ago the Tethys Ocean separated Asia from the Indian subcontinent. As tectonic plates shifted the two landmasses collided and the rocks on the ocean bed buckled. The fracturing of the Earth's crust released spurts of magma, which the advancing glaciers have transformed into magnificent granite peaks. Explorers who reach Everest and other iconic Himalayan summits are standing on mountains formed of marine limestone; fossils of sea creatures have been found in the sedimentary rocks. Although the Himalayas is one of the youngest mountain ranges on the planet, its network of snowy couloirs leads to many of the highest peaks on Earth, a lure for ambitious mountaineers.

But those who seek only the thrill of conquering these iconic peaks miss their deeper beauty. For centuries the Himalayas have inspired sages and writers, architects and artists. They have profoundly shaped the cultures of the Indian subcontinent, and are revered by the followers of five religions: Hinduism, Buddhism, Jainism, Sikhism and the indigenous Bon tradition of Tibet. One of the region's mountains has allegedly never been climbed, due to its religious significance. Mount Kailash basks in isolated splendour high on the Tibetan Plateau. It is considered to be at the centre of all things, since four great rivers flow out from its base:

the Indus, Sutlej, Brahmaputra and Karnali. The distinctive flat faces of Kailash, streaked with snow, are said to be formed from crystal, ruby, gold and lapis lazuli. In Hinduism it is known as the abode of Shiva – embodiment of enlightenment – and his spouse Parvati. It is here that the divine River Ganges cascades from Heaven to Earth and flows invisibly through the locks of Shiva's hair before gushing forth from a glacier.

In the past, mountaineers hoping to make an ascent of Kailish have been foiled by heavy snow. A Tibetan garpon from Ngari told the climber Herbert Tichy in 1936 that Kailish could only be climbed by someone who was 'entirely free of sin', adding: 'And he wouldn't have to actually scale the sheer walls of ice to do it – he'd just turn himself into a bird and fly to the summit.' The mountaineer Reinhold Messner was given the opportunity to climb Kailash by the Chinese government, having stunned the world by reaching the summit of Everest solo and without oxygen in 1980, but he declined. Messner said, 'If we conquer this mountain, then we conquer something in people's souls.'

For now, the only beings on Mount Kailash's peak are divine. From Shiva's position of aloof splendour on the summit, his third eye blazing with supernatural power and awareness, the lord of the mountain calmly surveys the swift triumphs and tragedies of human lives, the slow movements of the tectonic plates, the retreating glaciers and the drifting snow, the whole kaleidoscopic play of illusion that creates life in the world below.

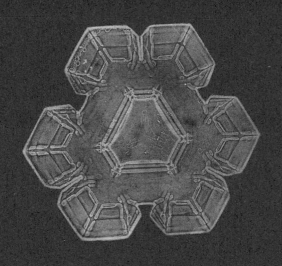

37.

Qasa

snow, ice

(Quechua)

'I believe that, since the memory of people, it has not been read of such a greatness as this road, made through deep valleys and high peaks, snow-covered mountains, marshes of water, live rock and beside furious rivers; in some parts it was flat and paved, on the slopes well made, by the mountains cleared, by the rocks excavated, by the rivers with walls, in the snows with steps and resting places; everywhere it was clean, swept, clear of debris, full of dwellings, warehouses for valuable goods, temples of the Sun, relay stations that were on this road.' So wrote Pedro Cieza de León in 1553 describing the Qhapaq Ñan, the Andean Road System that runs over 30,000 km along the length of the Andes, north to south from Ecuador to Chile. When the chronicler and conquistador encountered the Qhapaq Ñan it had been under construction and use for centuries. Meaning 'rich way' in Quechua (an indigenous language spoken by the Quechua peoples, the majority of whom live in the Peruvian, Bolivian and Ecuadorian Andes), the road connected communities across the Inca empire. Running messengers, or *chasquis*, carried news between relay stations; military manoeuvres took place, and goods of all kinds were transported on the backs of llamas.

The Qhapaq Ñan is a sophisticated feat of engineering through some of the world's toughest terrain. The Inca knew the landscape well and understood the force of the weather. They built steps into the steep slopes so that floodwaters would not erode them, and at high altitudes they paved the route with stone to protect its surface from the great snows, or *qasa*. The wonder expressed by Pedro Cieza de León at this organised way-making was reflected several centuries later, when in our own time the Qhapaq Ñan was inscribed as a UNESCO World Heritage Site, for being 'an exceptional and unique testimony to the Inca civilisation'.

Conversely, in Pedro Cieza de León's accounts of his fellow Spanish colonists, they are often confounded by the geography and weather conditions in the Andes. He describes: 'large mountains as covered by snow as others are by forests . . . [the Spanish] would go into this snow without knowing when or where they would end up . . . when they entered the snow-covered passes, huge snowflakes were falling from the clouds, which bothered them so much that they did not dare to raise their eyes to see the sky because the snow would burn their eyelashes.' The Incas were not daunted by the environment, but existed in respectful dialogue with it. They appeased it through sacred rituals such as Quyllurit'i ('bright white snow', from the Quechuan words *quyllu*, meaning 'bright white', and *rit'i*, 'snow'; sometimes translated as 'snow star') a festival that is still held annually in the Sinakara Valley of Peru. Quyllurit'i marks the disappearance of the Pleiades constellation at the end of April, and originally honoured Pariacaca, the god of water and torrential rains, although it has now been incorporated into the Catholic calendar. On the last night of the ritual, men known as *ukukus*, who dress as bears, climb high into the mountains to collect glacial ice for their communities. This ice is valued for its healing powers – although in recent years, with glaciers fast diminishing, the *ukukus* have elected not to take any of it away.

The ancient ice was not only healing but also pure and delicious. Merchants used to chisel blocks from the glacier, and brought them down along the Qhapaq Ñan to sell in the cities' markets, bundled up in plaited grasses to prevent them from melting. The ice trade is still evident in the name of one town, a staging post on the former ice route leading north from the snowy peaks of Pariacaca to the city of Lima – San José de Nieve Nieve, known for short as Nieve Nieve, or 'Snow Snow'.

38.

Barado

snow, hail
(Amharic: በረዶ በረዶ)

Is it true there is no Amharic word for snow? The consensus is that *barado* encompasses all forms of solid precipitation, from snow to hail; the word is even on record as having been used to refer to ice cubes. In polar cultures, the number of terms for specific kinds of snow is a clue to the cold climate; likewise, in Ethiopia the broad meaning of *barado* is an indicator of local meteorology.

For many years European explorers contested whether snow could be found in Ethiopia. Among the doubters was James Bruce, a Scot who travelled in Africa from 1769 to 1772. It was an era during which explorers were curious to locate the ghostly 'mountains of the moon' which ancient writers, including the astrologer and geographer Ptolemy, once believed fed the lakes at the source of the River Nile. (Today it is thought these might have been the Rwenzori, on the border of Uganda and the Democratic Republic of the Congo.) Bruce claimed there was no Amharic word for snow, and that snow 'was never seen' in Ethiopia. He observed hoar-frost on the grass of the Semien Mountains (ሰሜን ተራሮች) in northern Ethiopia, but no snow.

Yet if Bruce had encountered snow it would have been in these highlands – where green valleys are separated by high plateaus from which rise dramatic pinnacles of rock. The tallest peak in Ethiopia, Ras Dejen (4,550 metres), is among them. But rain – and thus snow, when temperatures drop low enough – is rare in these mountains in winter, because the majority of Ethiopia's precipitation occurs in the summer. Ras Dejen does sometimes experience strong snowfalls during the night, but since day and night temperatures vary greatly in the region (the temperature may be over 5° Celsius by midday), snow vanishes in a few hours.

However, to counter Bruce's denials, several texts do describe snow on the Semien Mountains. In one ancient inscription found on a stone chair at Adulis, a port on the Red Sea, an unnamed king from the third century CE describes a mountainous region in which he conducted his military campaigns as 'difficult of access and covered with snow, where the year is all winter with hailstorms, frosts and snows into which a man sinks knee-deep'. It is very likely this was the Semien. Later, snow was witnessed by the seventeenth-century Jesuit priest Jerónimo Lobo during his mission in Ethiopia. The explorer Henry Salt saw snow on the peaks on 9 April 1814, and records a fellow traveller, Nathaniel Pearce, who had experienced a heavy snowfall on the summit of Mount Hay a few years previously. Pearce told Salt that the snow 'did not come down with violence, but quietly descended in large flakes, like feathers'. Other nineteenth-century travellers remarked that the snow on the peaks of the Semien contrasted with the dark blue sky and the green vegetation at lower altitudes. There are even reports of two French officers, Ferret and Galinier, throwing snowballs at each other on Ras Dejen in January 1841 during their ascent to the summit – the first on record – and who could have a snowball fight without snow?

39.

Ttutqiksribvik

A place with a layer of snow on shore-fast ice, near an ice pressure ridge, where a boat and gear can be stored upside-down, and the sides of the boat banked with snow to protect gear under the boat. The bow of the boat is anchored to a piece of ice.

(Inupiaq, Kiŋikmiu or Wales dialect)

Ice crystals form on the ocean, just a greasy ooze at first, then shallow cakes, which accrete into a strong shelf that stretches from the shore out over the sea, to the cold border where the edge of the ice is lapped by the waves. Wind and tides crack the ice into floes that collide and buckle into ridges – the vertical elements of this otherwise flat landscape – that rise high above the ice shelf, and sink below the water sometimes reaching so deep that they anchor the ice in place on the ocean floor. Snowfall softens the jagged edges of this icefield as the winter passes, forming *qaimugut* or rolling snowdrifts, which resemble waves, and in the snow the tracks of polar bears appear then disappear, and humans approach and observe this temporary threshold as a good place to store a boat, an auspicious point for arrivals, departures and waiting – and it becomes: *tutqiksribvik*.

The seasonal cycles of snow and ice are central to the inhabitants of the small village of Wales, which sits on the westernmost outcrop of Alaska looking across the Bering Strait to the Chukchi Peninsula of Russia. Each spring, people leave their homes on the mainland and camp for weeks on the sea ice, taking shifts day and night to watch for the bowhead whales that pass on their migrations north to the Arctic. A boat is perched in readiness at the edge of the ice, harpoon and coiled line in the bow. When a darker shadow appears on the dark water at the horizon, or a fluke emerges, a puff of water from a blowhole, the other hunters emerge from shelter behind canvas windbreaks and launch the boat in pursuit.

In a culture in which the hunt is essential, not only to provide food for the community but also a focus for its spiritual well-being, a boat is a powerful visual symbol of a hunter's prowess and social standing. The *umiak* is a large boat that seats several people, traditionally made from a driftwood or whalebone frame pegged and lashed together, over which walrus or bearded seal skins are stretched. While the frame might be used by more than one generation of hunters, the skins must be continually restored.

The skins are also sourced during the spring hunt, from female walrus, who do not have as many scars as the cantankerous males; the hides must be protected until the hunting season is over and there is time to work on the boats again. The hunters in Wales take care the skins do not dry out by burying them in the snow.

Each spring *umiak*s are launched with solemn rituals – it was traditional for a graphite boundary to be drawn around the boat's waterline before it was set on the water, delineating realms of land and sea. Wood was a precious material in a treeless region, and seen as auspicious: sometimes an elderly woman scattered ashes gleaned from the shavings of newly carved paddles to create a path across the ice to the water's edge. Today, with the rolling waves of the Bering Strait as dangerous as ever, children still gather around the boat and crew to chant before they set off.

Much more than a boat, the *umiak* is also a place of shelter and even storage. Back on land after the hunting season is over, its upturned hull becomes a cave for storytelling and games, or a studio in which to whittle a fragment of bone from the catch into a child's toy. One writer recalls seeing large upside-down *umiak*s with rope swings strung from the frame, on which children played. In the past, boats were stored with amulets of dried year-old whale meat threaded beneath them for luck the following season.

Along the Arctic coastline many Inuit communities now use motorboats, but in Alaska *umiak*s remain popular. In place of oars or paddles, many modern boats include an outboard motor in the design; sometimes an aluminium frame replaces the skins. Along with adaptations to the vessels there have been other changes to the way the Kiŋikmiut hunters of Wales live. Winton Weyapuk Jr describes how snow machines have taken the place of dog teams and sleds, ATVs have become more popular than walking from house to house in a blizzard and, perhaps most importantly, many homes have a VHF radio in the kitchen, offering instant communication between hunters out on the ice edge and family back home, so there is less anxiety about the safety of the crew. This is a welcome development in a time when thinning sea ice and unpredictable weather patterns make the hunters' lives even more dangerous. It is more challenging now to camp on the ice edge, and difficult to harvest whales, seals and walrus as these marine mammals adapt their migration patterns to the capricious behaviour of the ice.

These technological changes brought with them new English words. Weyapuk grew up speaking Inupiaq with elders when out hunting, having begun to go out to sea from the age of eleven. The Inupiaq language taught him how to be safe in the boat and on the ice, then he had to speak English at school. Now English is the more commonly used language everywhere, on land, ice and sea; while the Wales dialect is not yet extinct, it is endangered. Fortunately, a new guidebook, the *Wales Inupiaq Sea Ice Dictionary*, has been compiled to preserve terms for snow and ice conditions, and the traditional knowledge enshrined in them. It is a characteristic of Eskimo-Aleut languages such as Inupiaq and Greenlandic that they are polysynthetic, meaning that they pack a great deal of information into a single word, through stringing together roots and suffixes, a little like the hunter's equipment stored under an upside-down boat. While ominous entries such as '*Mizagluk*: water on the ice, often covered by snow; a dangerous spot to step on' are becoming tragically ever more relevant, terms such as *tutqiksribvik* prove that snow and ice are still the working materials of this community, despite the dangers, and whether a boat is made of aluminium sheeting or walrus hide.

40.

Ais i pundaun olsem kapok

'the ice it falls down like cotton', ice-cotton, snow
(Tok Pisin)

On the slopes of the mountain range that crosses New Guinea from east to west grows the world's third largest rainforest. The eyes of international conservationists are trained on this highland region, which holds thousands of rare species: giant ferns explode from the rich soil and tiny mosses race up the trunks of palm trees into the canopy; between the weeping figs and sweet-smelling acacia the undergrowth is starred with the bright flowers of orchids.

There are more varieties of orchid here than anywhere else on the planet. The island has preserved many ancient plant species from a 'super-continent' known as Gondwanaland, which covered one-fifth of the Earth's surface 100 million years ago. Over aeons this landmass began to shift and fragment, taking on the form of

the Southern Hemisphere as we know it. When the Australia–
New Guinea continental plate broke off from Gondwanaland,
it remained isolated and maintained a unique ecosystem. By the
time New Guinea reached its current position in the western
Pacific Ocean about 10 million years ago, the island had preserved
many of the ancient plant species of Gondwanaland, growing
alongside newer varieties that had migrated from the shores
of Asia.

Snow is extremely rare in New Guinea's tropical climate,
and even hail (*ren ais*, or 'rain-ice' in Tok Pisin, the creole language that is used throughout New Guinea, although 1,100 other
languages are also spoken) is uncommon. Any unusual weather
events take on dramatic significance and are remembered for generations. When the geographer Ronald Skeldon met the Guhaku
people in the 1970s in the Eastern Highlands, they told him the
story of an epic hailstorm around the close of the nineteenth
century. (This might have been caused by the eruption of the
Indonesian volcano Krakatoa in 1883, which triggered a sequence
of weird weather events around the world for several years after-
wards, including powerful blizzards and record snowstorms in
unexpected locations; the chaos was the equivalent of that pro-
voked by Iceland's Eyjafjallajökull in our own times.) The Guhaku
witnessed an earthquake, which brought down the branches of
trees, made the thatched roofs of their houses collapse and even
started landslides. The atmosphere became so dark and cold that

they had to carry fire-torches as they fled from their disintegrating homes. The next day the ground was covered with ice. But it did not stay long. As soon as the sun shone again it melted away, revealing the devastation to crops such as cassava and sweet potato. The Guhaku people sacrificed white pigs in response to this 'white' weather – a spiritual offering, but perhaps now also an essential source of food.

The Guhaku emphasised the damage to trees and crops in their accounts of the storm. The word they used for hail – *ovisopa* – was in fact the term for a small white flower of the *Pipturus* species (a kind of nettle), which was associated with hailstones because of its colour and shape. In their wonder at this unusual precipitation falling from the heavens, the islanders chose to describe it by comparison to plant life, growing up from the earth – something much more familiar to them, and equally temporal. And this is true also of the term used more widely for snow on New Guinea: 'the ice it falls down like cotton' or 'ais i pundaun olsem kapok' (*kapok* is the word for a cotton plant; *olsem* is used in place of words that perform linkages: 'like', 'in this way'; *pundaun* means 'fall down'). The white fibre spheres that surround the seeds of the cotton plant, which has long been cultivated on the island, do indeed resemble snow – especially once the seeds are ripened, and burst from the bolls in a soft blizzard across the verdant landscape.

41.

Hagelslag

hail burst, chocolate hail
(Dutch)

Some people do not understand how to take holidays. My own family went abroad once every five years or so, always to the Netherlands, so that my father, an art historian who studied old Dutch paintings, could pursue his research in the Rijksmuseum. It always seemed to be winter. We spent most of our time indoors, usually within the dim Rijksmuseum galleries. I must have been shown the many paintings that celebrate the Dutch waterways; the famous scenes, such as Hendrick Avercamp's *Winter Landscape with Skaters*, with its crowds of richly dressed couples and fishermen and old women and young children skating and sliding on an icy river – but I do not remember them. I remember an acquaintance of my father's who gave me an ostrich feather dyed bright orange, and the frosty darkness in a city square where I built a snowman one evening, and the lights of the tall houses reflected in the still canal water. And I developed a taste for *hagelslag*.

Hagelslag, or chocolate hail, is a condiment eaten as a topping on bread year-round, often at breakfast time. It is properly pronounced with two harsh guttural gs which sound a little like the rumblings of a hungry stomach. The sweet sprinkles can be found in over twenty different varieties: chocolate *hagelslag* (milk, white and dark), extra-large sprinkles, chocolate flakes (*chocoladevlokken*), and even forest-fruit flavoured sprinkles. Low-sugar versions are available. Supermarkets have not just shelves but whole sections dedicated to *hagelslag* (both own-brand and commercial) in bright cardboard boxes and even, sometimes, dispensers. Over 750,000 slices of bread topped with *hagelslag* are eaten every day in the Netherlands, and 14 million kilos of *hagelslag* are consumed annually.

Hagelslag is not a new craze – it is over a century old. One bleak autumn day in 1919, B. E. Dieperink, the director of the liquorice company VENCO, was watching *snagel* or 'snowy hail' fall and bounce off the cobbled pavement outside his office. He had the idea of making white aniseed-flavoured sprinkles to look like hail, and using them as a topping on bread. In that austere era the sweet, brittle *hagelslag* became extremely popular; VENCO patented the name, but other companies such as De Ruijter were already making the little sprinkles. De Ruijter named their lemon, raspberry, and aniseed sprinkles *Vruchtenhagel* ('fruit hail'), while their chocolate varieties are *chocoladehagel* ('chocolate hail'). An orange flavour was introduced for the birth of Princess Beatrix (of the House of Orange-Nassau) on 31 January 1938. These sprinkles may be tiny compared to the earthworks by which the landscape of the Netherlands is redeemed from the sea, but they are as much a Dutch institution as skating on rivers in winter.

42.

Eira

snow
(Welsh)

When the Welsh mountains are blanketed in snow in midwinter, it is high summer 13,000 km away on the Patagonian Steppe. Here, between the Andean glaciers and Atlantic coast, a brave group of Welsh emigrants established the first permanent settlement south of the Río Negro during the nineteenth century. After a two-month journey from Liverpool, the converted tea clipper *Mimosa* landed in Patagonia in July 1865 with 153 Welsh settlers aboard, drawn by the prospect of land of their own, and more freedom than they could hope for in Wales. A generation earlier, an education act had made the use of the Welsh language in schools illegal. The British government's justification for this ban is shocking to read today: 'The Welsh language is a vast drawback to Wales, and a manifold barrier to the moral progress and commercial prosperity of the people . . . It dissevers the people from intercourse which would greatly advance their civilisation, and bars the access of improving knowledge to their minds.'

The Argentine government offered the Welsh settlers the Lower Chubut Valley, arid land that was much harder to work than the fertile valleys back home. One of the first sermons preached in chapel after the pioneers' arrival was on the subject of 'Israel in the Wilderness'. Their attitude towards their adopted environment was influenced by these Bible stories about the desert as a region hostile to human life, a place of solitude and bleak despair, but ultimately an arena of trial for the faithful. As time passed, they realised, largely thanks to conversations with the indigenous Tehuelches, that there might be better land out there – and they set off on expeditions to find a fabled mountain zone of lush vegetation and snowy peaks to the west.

The challenges of exploring Patagonia can be inferred from the names the settlers gave to the regions through which they passed: Hirlam Uffernol ('Hellish Journey') and Banau Beiddio ('Daring Peaks'). The promised land proved elusive. Nearly two decades on from *Mimosa*'s landing, all the irrigable land on the coast had been claimed. More territory was needed. Finally, in November 1885, explorers reached the Andean foothills, which they named Cwm Hyfryd or 'Pleasant Valley'.

'We camped that night on the banks of a small stream that sprang from the mountain behind us,' wrote Thomas John

Murray, 'and on its banks grow blackcurrants, redcurrants, rhubarb, raspberries, watercress, birch and other trees.' The beautiful scenery evoked fond memories of the Welsh hills, after years on the plains. Another pioneer remarked: 'the panoramic view of the Andes . . . opened up before us to the right and left. It was one of the grandest sights that it is possible for a man to see.'

Until the third generation, the settlers were able to pursue their goal of speaking Welsh. They might even have had cause to use the word for snow, *eira*, on glimpsing the Andean peaks. (It is pronounced to rhyme with Sarah, with a rolled *r*.) But from the 1930s onwards the language was restricted to home and chapel after the Argentine government's attitude to a multilingual nation shifted; then, during a military dictatorship in the 1970s and 1980s, people were banned from giving their children Welsh names. Speaking Welsh in the home grew more infrequent. Ironically, as government policies in the UK became more inclusive, the Welsh Patagonians found their freedoms restricted.

Since the centenary of settlement in 1965 there have been efforts to encourage the use and enjoyment of the Welsh language once more. Schools often teach in both Welsh and Spanish. The modern Patagonian Welsh dialect is distinct from the Welsh spoken in the UK, as it incorporates Spanish loan words and has been influenced by Spanish pronunciation too. Although these two nations are at opposite ends of the globe, love of language and landscape still link their stories.

43.

Itztlacoliuhqui

God of Frost
(Nahuatl)

Mexico may be known for its hot desert regions, but snow does fall frequently in highland areas such as Toluca and Durango, and on the exposed summits of Citlaltepetl, Popocatepetl and Iztaccíhuatl. One of the highest mountains in Mexico, Iztaccíhuatl means 'white woman' in the Nahuatl language – an apt description for the four peaks that resemble the head, torso, knees and feet of a giant sleeping figure covered in a blanket of snow. Before the Spanish conquest of the Aztec city of Tenochtitlan in 1521, these slopes were scaled annually by a lone figure dressed as Itztlacoliuhqui, the god of frost. When the pilgrim reached the summit of Iztaccíhuatl, he removed the distinctive mask that gave him the look of the deity, and left it there – and with it, the perilous cold . . . The return of Frost to his home on the mountain ensured that he would not harm the harvests of maize, beans, tomatoes and chiles growing at lower altitudes.

According to Aztec mythology, this god of winter was born out of the arrogance of Tonatiuh, the god of the sun, who was demanding obedience from all the other gods. Enraged by such arrogance, Tlahuizcalpantecuhtli, the god of dawn, the morning star and the planet Venus, fired an arrow at him. The shot missed, and the Sun retaliated – piercing the god of dawn through the head. At this moment, he was transformed into a new god, Itztlacoliuhqui, and thenceforth associated with the night, and with the cold north, and lifelessness. Itztlacoliuhqui was a busy deity: his realm also covered ice, winter, sin, punishment and human misery. His name is usually translated into English as 'curved obsidian blade', but Nahuatl scholar J. Richard Andrews notes that it could also mean 'everything has become bent by means of coldness' or 'plant-killer-frost'.

Only a few images of Itztlacoliuhqui survive in old manuscripts. He is usually depicted wearing tasselled robes in a patchwork of pale violet and yellow hues, with variegated stripes. His headdress is rent by a bloody representation of the arrow that changed his nature from hope to despair, and his face is obscured by a piece of finely curved black obsidian. Some say this 'blindfold' reflects his impartial role in dispensing justice, others his imperviousness to the hardship poor farmers and their families suffered in cold weather. Either way, the black volcanic glass is a fitting motif – frost cuts as sharply as an obsidian blade.

The Florentine Codex, created in the sixteenth century by the Spanish Franciscan friar Bernardino de Sahagún in collaboration with indigenous advisers, explains the role of cold in Aztec society: 'Once yearly the cold came. During the feast of Ochpaniztli the cold began. And for 120 days this persisted and there was cold.' The duty of all Aztecs during the cold season was cleanliness; the name of the eleventh month of the Aztec calendar, *Ochpaniztli*, means 'sweeping'. Itztlacoliuhqui carries a straw broom, symbolising his function as the sweeper away of the old. For frost wasn't always a dreaded phenomenon. It was a sign that the seasons were turning and the time of sowing, and then harvest, would come again. Itztlacoliuhqui's devotees were serious about the sweeping: not merely houses but also the streets and hillsides were swept clean, daily, during *Ochpaniztli* – making way for a fresh start and new life.

44.

Pana

snow knife

(Inuktitut: ◁ᓇ)

How do you transcribe a language that has never been written down before? This question preoccupied French and English missionaries in the polar north when they encountered the Inuktitut language, the name of which derives from the words *inuk* (meaning 'person') and *-titut* ('in the manner of'). As in: this language is the way people speak. After years of debate, in the 1870s the missionaries adapted the Cree script to forge a syllabic writing system; the words in the Inuktitut dictionary look very different from our own. The syllabics are known as *Qaniujaaqpait* (ᖃᓂᐅᔭᐊᖅᐸᐃᑦ), which derives from the root word *qaniq* meaning 'mouth'; it is opposed to *Qaliujaaqpait* (ᖃᓕᐅᔭᐊᖅᐸᐃᑦ) or Latin script (derived from *qaliit*, a word describing the markings or the grain in rocks). This division between breath and mark, between

Inuit and Western writing, highlights the differences between the oral and written traditions, and that which is temporal and that which is set in stone.

It is a division that underscores many tensions in the region where Inuktitut is spoken: north of the treeline in Canada, from Newfoundland and Labrador on the east coast to parts of Manitoba, as well as across the state of Nunavut, the most northerly and largest in Canada – which stretches from Hudson Bay in the south to the archipelagos of the Arctic Ocean. Nunavut is also the newest state, having been divided from the Northwest Territories on 1 April 1999, a historic act in recognition of traditional Inuit lands.

One of the first written works of Inuktitut literature is the novel *Saanaq*, which actually began as a dictionary. Mitiarjuk Nappaaluk (1931–2007) began writing it in the 1950s when Catholic missionaries asked her to make a phrasebook of common words from everyday life in syllabics. (During the writing, which took almost twenty years, Nappaaluk met anthropologist Bernard Saladin D'Anglure, who encouraged her work on the novel and served as her translator.) In *Sanaaq*, Nappaaluk tells the story of an extended Inuit family and the various activities – such as making and repairing clothing, gathering bird eggs, and hunting seals – that make up their existence, living almost entirely off the land. The building of shelters is a recurring theme.

An essential skill for survival in the Arctic was once the

ability to build a house using nothing but snow. Two people working together could construct an *iglu* in less than an hour, and their swift work made the difference between life and death in bad weather conditions. Snow from one storm is often used to build a shelter from the next. An *iglu* requires consistent snow: blocks with softer, looser streaks would crumble during handling. The snow must be hard, yet not so hard-packed that it cannot be cut into blocks. A drift of dry snow laid down in a single storm is ideal. The importance of 'good snow' is described in *Sanaaq*:

Qalingu went out for some good snow. In his hand was a snow knife and on his arms airqavaak [long-sleeved gloves]. As for Sanaaq, she had been too busy yesterday to do all she had wanted to do.

Arnatuinnaq told her, 'I'm going to stop up the cracks on the outside of the igloo. [. . .]'

Qalingu dug a hole in the snow, but it was not good snow. He said, 'It isn't any good, so I'll make the igloo out of packed snow . . . We'll trample it today to pack it together. It will harden overnight.'

He cut out a large number of blocks that he broke up with his snow knife. Arnatuinnaq then used her feet to pack the snow. Qalingu told her, 'We're going to be very cold tonight. I probably won't be able to start building our igloo before tomorrow.'

'The snow is very powdery. It will take long to harden,' replied Arnatuinnaq. 'There's some wet snow a bit further away . . .'

Several episodes in Nappaaluk's novel remind the reader that the right snow is no use without the right tools. The *pana* (<ᴄ) or 'snow knife' is one of the Arctic hunter's most valued possessions. The blades of the earliest *pana* were carved from polished bone, with an ivory or antler handle, but after contact with explorers, metal knives were often used. It is sometimes claimed that bone was better, because it is able to transmit the sensation of the snow to the hand holding the knife. Whatever the material, the blade should be long and sharp enough to slice large blocks out from the snowbank. This area may become the base of the *iglu*, with the pleasing result that the whole structure is built entirely from snow taken from within the space occupied by its walls – hence the floor is much lower than the snow outside. The snow blocks (each about 4 inches thick) are neatly placed in a circle; then the *pana* is used to trim the top edge of the blocks into a sloping upwards curve; it looks like the diminishing spiral of a snail shell, as opposed to a brick house where the bricks are all on one level. The second layer of blocks is laid on top, with a gentle inward curve, which is again formed by chiselling away at the blocks with the *pana*. This inward lean becomes more acute as the walls grow higher. The bevelled edges allow each block to gain support from the previous one, and from the one below. When the last block is inserted in the apex of the roof, it must perfectly fit this tensile structure or the whole *iglu* could collapse.

When two people are working as a team, one will build the snow blocks up from the inside as the shelter rises around them. The final action is to chisel a round, low doorway through which they can emerge to admire their efforts. Then the *pana* is triumphantly stuck in the snow, which doubles as a useful toolrack, until it is needed again.

Although made of such subtle and shifting material, these domes are wonderfully sturdy structures: the snow provides a good mortar as well as bricks. Where light can be seen through the cracks, a little more snow is added then swiftly rubbed in with the flat blade of the *pana* (known as 'chinking'). Once people settle inside, the warmth of bodies combined with the cold wind outside will create a layer of ice to 'set' the snow. A place to find shelter and rest, a microcosm of the wider community, the *iglu* is also where stories are told in the evenings: hunting histories and founding myths passed on by word of mouth. The past becomes the present here. A knife can sever but it can also build.

45.

Jäätee

ice road
(Estonian)

In wintertime roads appear over the frozen surface of seas, lakes and rivers in Estonia. On these freeways, there is nothing but whiteness as far as the horizon, perhaps a fringe of dark conifers in the rear-view mirror, and total silence. This makes for an eerie driving experience on the route that runs for 26 km from the port of Rohuküla on the mainland coast of Estonia across the Baltic Sea to the island of Hiiumaa. Here, on Europe's longest *jäätee*, there are few markings to follow. A car spins along in the tracks made by previous vehicles, passing an occasional traffic sign or guided by large juniper branches, 'planted' upright like trees growing magically from the snow, to indicate the edges of the road. During the journey, which can take at least an hour, the traveller watches the cinematic view play out through the windscreen, and it might feel as if they are driving off the edge of the world. A good soundtrack is recommended. It is both a relief and a disappointment when Hiiumaa's coastline appears, a faint shadow in the far distance. The meditative skyline and the slim silver *jäätee* are soon replaced by salted tarmac and slushy roundabouts, and all the messy dockside infrastructure of tollgates and Portakabins, streetlamps and telegraph poles.

Ice roads are open only during the hours of daylight, and even then snow flurries may decrease visibility. The road is a shortcut, as the crow flies, but no one should rush it – or slip into a dawdling dream. Drivers must keep to speeds of between 25 km/h and 40 km/h – the lower limit is important. No stopping is allowed. This is a precaution against changes to the car's rate of progress causing a wave under the ice; if such a wave accumulates it can be strong enough to crack it. For the same reason vehicles must travel at least two minutes apart, and so drivers wait at the shore for a green light before they set out. These strict safety measures are accompanied by an unexpected road rule: it's forbidden to wear seatbelts, because drivers and passengers might have to exit the car speedily in the event of the ice cracking.

But such terrifying incidents are rare. The ice must be at least 22 cm thick before the *jäätee* is opened to traffic. As a result, the road is unlikely to stay open all winter, and in 2015 it did not open at all. Estonia ranks among the smallest countries in the world, and keeping a connection between the mainland and the many islands is important, especially when the ferries that chart this ice-free route in summer are enjoying a long hibernation.

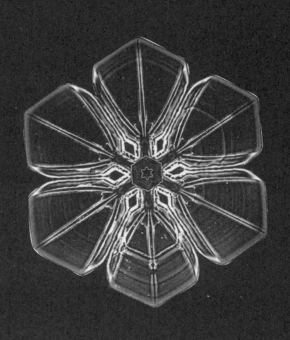

46.

Sparrow batch

spring snow
(Newfoundland English)

Newfoundland has long been a location for significant arrivals and departures. At the most easterly tip of Canada, this massive island juts into the wild Atlantic Ocean, and its cliffs, even though rugged and daunting, have been a welcome sight to those who have sailed westwards for many days. In the eleventh century the Norse explorer Leif Eriksson landed here – becoming the first known European to set foot on the North American continent, several centuries before Christopher Columbus. (The land was already inhabited by indigenous people of the Beothuk culture.) On disembarking, Eriksson encountered a mild climate, salmon in the rivers and plentiful grapevines, and so he named the fertile island 'Vinland'.

While Leif Eriksson might have found the weather here more clement than on his home farm in Greenland, the Newfoundland winters are harsh by most standards. The shoreline freezes over with drift ice, and the central rocky plains (forged by glaciers in former times) are hidden under deep drifts of snow. As the weather grows more temperate in March, the most common form of precipitation is 'silver thaw', a freezing rain or drizzle that covers everything in ice. 'Silver thaw' is just one of many evocative weather terms in an island dialect. 'Newfinese', as it is known among its speakers, is derived from emigrants from the West Country – Bristol, Devon and Cornwall – who followed after Eriksson, and it differs markedly from the English spoken elsewhere in Canada and the North Atlantic. One phrase you're likely to hear just before spring is 'sparrow batch'. A batch is a heavy, substantial fall of snow; a sparrow batch is a late snowfall in April, said to bring back the sparrows.

On this island, which is affectionately known by locals as 'The Rock', bird migration is a crucial sign of seasonal change. The species that might be seen hopping over the batch are the white-throated sparrow (with its distinctive white and yellow crown stripes), the swamp sparrow, the fox sparrow, the savannah sparrow and Lincoln's sparrow. In late autumn they flock together to hunt for seeds among the overgrowth. But the seed supply is finite, and none of these tiny birds can remain in a place in which winter temperatures will drop well into the minuses. They must spend the darkest months of the year elsewhere. As the nights lengthen, they follow an instinct in their genes going back many generations, and set their course south for wintering grounds in the mid or southern United States. Their flyway covers a distance of thousands of miles, and requires navigational skills that Leif Eriksson himself might have envied. All winter, the sparrows fatten up for the difficult return journey, when the last of the snow calls them back north to breed.

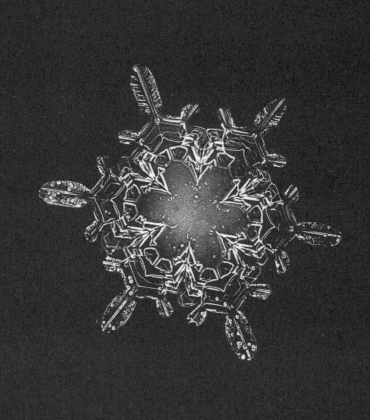

47.

Hau kea

white snow, snow
(Hawaiian)

In this tropical archipelago, snow is most likely to be found on a simmering crater. In winter the temperature at the summits of Mauna Loa, Haleakala and Mauna Kea – the state's three tallest volcanoes – drops to below freezing. Mauna Kea means 'white mountain' and it is a sacred site, the home of one of the four snow goddesses, Poli'ahu – the most beautiful Hawaiian goddess and sworn enemy of Pele, the fiery goddess of Volcanoes.

The peak of Mauna Kea is considered the highest island mountain in the world. Up there, you're closer to the planets, the sun and the stars, the rising and setting of which were once used by Polynesian navigators to chart the ocean's vast, formless waters, and thus brought the first humans to these islands. With this ancestral connection to the heavens it's no surprise that today the

University of Hawaii has a world-class astronomy facility. Many of the observatory instruments, thirteen telescopes among them, are located on the summit of Mauna Kea. In recent years, while astronomers are looking up into the skies through sophisticated lenses, other eyes have been on the activity on the peak. In 2014 Mauna Kea was selected as the proposed location for another instrument – the giant Thirty Meter Telescope. Leading academics embraced the idea of more technology on the mountain, seeing it as a natural continuation of the astronomical tradition of the early navigators. The act of placing one of the world's largest segmented mirror telescopes on Mauna Kea was to them an inevitable combination of culture and science, and a tribute to the ongoing human quest for knowledge of distant worlds.

But not everyone sees the development as respectful of ancient knowledge. Many believe that the mountain is sacred and do not wish to see the site developed further for science. Environmentalists are concerned about the displacement of rare native bird populations. In 2015 demonstrations began against construction, with the activists blockading the road to keep building crews off the summit. They see themselves not as protestors but as the mountain's protectors, or *kia'i*. They have bravely battled arrests, and, during the winter months, cold weather. Sturdier tents crept into the protest camp, space heaters and stores of thermal clothing. Then, in January 2020 came a winter storm so severe that around two feet of snow dropped on Mauna Kea. The snow, which drifted up to 8 feet deep in places, closed public road access to the summit. Nature or perhaps Poli'ahu doing the work the *kia'i* had done for years. That winter, the two sides contesting the soul of Mauna Kea agreed to a two-month truce: the consortium that wanted to build the telescope would not start construction, and the *kia'i* took a break from life in camp conditions.

Protest songs were sung in Hawaiian, a richly polysemous language in which a word or phrase can have many coexisting meanings. *Hau* means snow, but the term can also mean anything cold (ice, frost, dew); a cool breeze; a type of mammoth; a hibiscus tree with heart-shaped leaves (*Hibiscus tiliaceus*) which grows in warm places; mother of pearl; or a pumice stone. Only the term *hau kea* ('snow white') exclusively denotes what we think of as 'snow'. This expansive approach to the material world can also be found in the writings of Johannes Kepler, the seventeenth-century German scientist who investigated the vast reaches of the solar system, and whose studies are cornerstones of Western astronomy. He made a connection between huge masses of burning hydrogen and the microscopic forms of frozen crystals. After charting the elliptical movements of the planets, Kepler moved on to investigate snow, writing that it too 'falls from the heavens and looks like the stars'. Perhaps instead of looking through the Thirty Metre Telescope, humans need only to climb the peak of Mauna Kea on a winter evening – singing a song of solidarity and celebration – to understand the stars.

48.

Virgen de las Nieves

The Lady of the Snows
(Spanish, worldwide)

In the fourth century, during the rule of Pope Liberius, a pious Roman couple wished to donate their wealth to the Church – but in what form should they do so? Their dilemma was solved when the Virgin Mary appeared to them, entreating them to build a chapel dedicated to her wherever and whenever she gave them a sign. One August evening, after a blazing hot day, a light scattering of snow fell on the summit of the Esquiline Hill in Rome. The rich couple were sure it was the miracle they were waiting for, and they ordered a basilica to be built on the spot. Many centuries later, each year on 5 August at the conclusion of the Solemn Mass in the chapel (still popularly known as the Basilica of Our Lady of the Snows) a shower of white rose petals is dropped from the dome, and at sunset, another artificial snowfall is staged in the square outside. Although the Catholic Church believes the story to be a legend, not a true miracle, the liturgical feast of the Virgin of the Snows is celebrated worldwide, and her appearance has inspired many more church buildings.

On 5 August 1717, Martín de Mérida was walking to Granada across the Sierra Nevada, when on the summit of Carihuela he was surprised by a fierce snowstorm. These peaks and precipices can be perilous, even in summer. Facing the prospect of freezing to death in the mountains, he had no hope but to pray for rescue. The Virgin Mary appeared in the heart of the storm and calmed the elements, then showed him which direction he should take to the safety of the valley below. Ever since, these frozen cliffs have been known as Tajos de la Virgen, and the Virgen de las Nieves is considered the patron saint of Sierra Nevada. The following year the grateful priest built a hermitage, which (not surprisingly) was soon destroyed by weather. In 1724 its ruins were replaced by another, located a little lower, in the Prados del Borreguil; but this one couldn't resist the attack of the elements either. In 1745 a third hermitage was built at a still lower altitude. It has survived. Today it houses a nature classroom for the Sierra Nevada Natural Park, but its origins are recalled by its name, the Old Hermitage. Each year pilgrims climb to a different summit on the range to hold a mountain mass for the Virgen de las Nieves. The only thing more miraculous than snow in summer, it seems, is shelter from the storm in a time of need.

Closer to the equator, the Virgen de las Nieves often appears in the Canary Islands, where she is the patron saint of La Palma and Lanzarote. The fervour with which the inhabitants of these balmy islands in the Atlantic Ocean worship her is fuelled by the belief that she defends them from pirate attacks, volcanic eruptions and drought. *Año de nieves, año de bienes*, goes old Spanish weather lore: 'Year of snow, year of goods.' Snow that lies late on cultivated soil brings welcome moisture for crops in hot regions. A terracotta figure of the Virgin is kept in a sanctuary on La Palma, and during the Bajada de la Virgen, which has been held every five years since a drought in 1680, pilgrims carry her on a silver throne to the capital, Santa Cruz de la Palma. In Tenerife, the largest of the Canary Islands, a small hermitage dedicated to the Virgin perches on the slopes of Mount Teide. Ermita de las Nieves is the highest Christian temple in all Spain – its cool interior a dramatic contrast to the azure skies and surrounding lava-strewn slopes. The Virgen de las Nieves crossed the Atlantic with five immigrants from La Palma in the 1960s, who made their new homes in the town of Cagua in Venezuela. Shortly after their arrival, her devotees began to raise funds to organise an exact replica of an icon from their homeland – a sumptuous painted carving of the Virgin in a white mantle surrounded by stars – which was at last enthroned in the church of San José de Cagua in 1976 – a very long way from the Esquiline Hill.

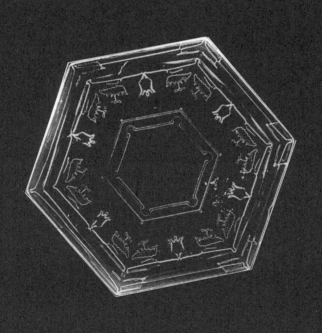

49.

Zud

severe winter

(Mongolian: зүд)

On the Mongolian Steppes, wealth is counted by animal hooves. For centuries nomadic herders have guided their livestock between the lush pasture of the lowlands in spring, and the more sheltered highlands in winter, where they can escape the punishing wind. This grassland region, so sparsely populated by humans (Mongolia is home to a mere 3 million people), is not short of other animals. A herder will typically own 1,000 sheep, as well as goats, cattle, horses or camels. Pigs, poultry and bees are sometimes kept too, and herders supplement their living by selling milk and wool, and animal hides. One third of Mongolia's population depends entirely on farming for its livelihood, but that figure is diminishing, along with the herder's wealth.

Mongolia is known by its inhabitants as the 'Land of the Eternal Blue Sky' or *Mönkh khökh tengeriin oron*. Although the steppes experience over 250 sunny days a year, they can be bitterly cold when a front of heavy air blows south over the mountains from Siberia. And even blue skies bring their own anxieties. It is well known among herders that after a hot, dry summer, they can expect to endure an especially bitter winter. Too much snow or too little snow will cause a natural disaster unique to Mongolia: the *zud* or *dzud*, a severe winter in which many animals die either from starvation or the cold. The *zuds* can arrive in different ways, and are classified accordingly:

- The *tsagaan* (white) *zud* is caused by deep snowfall that prevents animals from reaching the grass.
- The *khar* (black) *zud* is caused by a lack of snow in grazing areas, often in the Gobi Desert region, leading both animals and their owners to suffer a shortage of water.
- The *tumer* (iron) *zud* is caused when a brief period of mild weather is followed by a return to sub-freezing temperatures. The snow melts and then freezes again, creating ice-cover that prevents animals from grazing.
- The *khuiten* (cold) *zud* occurs when the temperature is very low for several days. The cold, combined with strong winds, prevents livestock from grazing; the animals use most of their energy to keep warm.

The *zud* has always been dreaded by the herders, but around the turn of the millennium a series of very severe *zuds* caused an unprecedented disaster, with the loss of millions of animals. Herders did not always manage to salvage the carcasses of their livestock, which lay on the steppe, where golden eagles and other scavengers picked their bones. Desperate men slaughtered their own stud animals, in the hope that sheep that did not lamb would be stronger, and more likely to survive the winter. It was not only the sheep facing starvation – the herders did too.

The economic crisis that followed forced many herders to abandon their traditional pastoral existence and move to urban areas, in particular the capital Ulaanbaatar. Located in the Tuul river valley, this city is one of the world's coldest, experiencing temperatures as low as minus 30° Celsius in January. The population of the capital increased by a quarter at the end of the millennium, as herders became part of the unemployed and poor in the growing northern outskirts of town. The *zud* has the power to change the shape of a whole society.

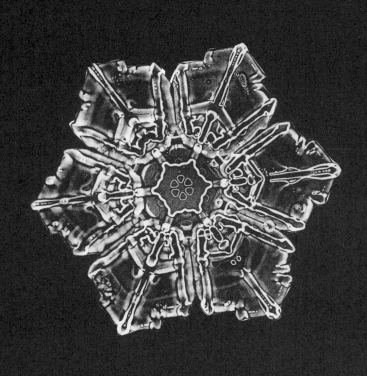

50.

Suncups

hollows in melting snow surfaces
(English)

Shallow hollows sometimes form in snowfields, as tiny as a watch face or larger than the dial of a grandfather clock. This honeycomb pattern is created when a light breeze along the surface of the snow encourages evaporation from the ridges, while the sun radiates in the groove between them, melting the depressions. Since the evaporation of snow demands more heat than does melting, less snow is lost from the high points in the form of vapour than is lost from the hollows in the form of meltwater. The hollows melt faster than the points evaporate, growing deeper and deeper. In the Northern Hemisphere suncups form more swiftly on the southern part of a snowfield, where it first catches the sun – and so the pattern gradually migrates northward, as if lured by colder places.

Acknowledgements

I am grateful to the following individuals who generously shared with me their linguistic knowledge and their enthusiasm for snow: Theophilus Kwek (Chinese), Hanne Busck-Nielsen, Mette-Sophie D. Ambeck (Danish), Carinne Piekema, Natasha Herman and Adrian Kruit (Dutch), Mikhel and Jenny Zilmer (Estonian), Mark Olival-Bartley (Hawaiian), Matthew Teller and Marryam Reshii (Kashmiri), Anna Iltnere (Latvian), Egidija Čiricaitė (Lithuanian), Gregory Cowan (Mongolian), Marlene Creates (Newfoundland English), Imi Maufe (Norwegian), Laura Fernández-González (Spanish and Quechua), Elspeth Napier, Ken Cox of Glendoick Garden Centre (Tibetan), Jenny Gal Or (Tok Pisin), Ralph Kiggell (Thai), Nasim Marie Jafry (Urdu), and Phil Owen (Welsh). To these friends, and many others whose acts of kindness sustained me through a long and difficult winter, my warmest thanks.

References

Note: All country names are given as they are recognised today.

Prologue

Italo Calvino, *Marcovaldo: Or, the Seasons in the City*, trans. W. Weaver (London: Vintage, 2001).

Seamus Heaney, 'Digging', from *Death of a Naturalist*, (London: Faber and Faber, 1966), reprinted by permission of Faber and Faber Ltd.

For more on the 'Inuit Snow Conspiracy' see Laura Martin, '"Eskimo words for snow": a case study in the genesis and decay of an anthropological example', *American Anthropologist* 89, 2 (June 1986): 418–23, and Steven Pinker, *The Language Instinct: The New Science of Language and Mind* (London: Penguin, 1994).

1. Seaŋáš, Sámi

For the international snow classification ('The International Classification for Seasonal Snow on the Ground') see:

https://www.hydrology.nl/ihppublications/178-the-international-classification-for-seasonal-snow-on-the-ground.html.

2. Yuki-onna, Japanese

This is a retelling of an original tale by Lafcadio Hearn, which can
be found in *American Fantastic Tales: Terror and the Uncanny from
Poe to the Pulps* (New York: Library of America, 2009) pp. 282–5;
published originally in *Kwaidan: Stories and Studies of Strange Things*
(New York: Houghton Mifflin, 1904).

3. Immiaq, Greenlandic

Immiaq recipe taken from Keld Hansen, *Nuussuarmiut – hunting
families on the big headland: demography, subsistence and material
culture in Nuussuaq, Upernavik, Northwest Greenland*, Man & Society
Series, no. 35 (Copenhagen: Museum Tusculanum Press, 2008).

4. Smoor, Scots

James Hogg, *The Shepherd's Guide, Being a Practical Treatise on the Diseases
of Sheep, Their Causes, and the Best Means of Preventing Them; with
Observations on the Most Suitable Farm-stocking for the Various Climates
of this Country* (Edinburgh: Archibald Constable, 1807) and *The
Shepherd's Calendar* (Edinburgh: William Blackwood, 1829). Readers
new to Hogg's work are advised to start with his autobiography,
The Private Memoirs and Confessions of a Justified Sinner (Edinburgh:
Canongate Canons, 2018).

Robert Burns, 'Tam o'Shanter', in *The Complete Illustrated Poems, Songs and Ballads of Robert Burns* (London: Lomond, 1990).

8. Sheleg, Hebrew
The quotations are taken from the King James Bible.

9. Sastrugi, Russian
Finnesko: a soft hide boot, originally from Lapland, ideal for travel in cold climates.

'annoying obstacles . . .' Ernest Shackleton, *The Heart of the Antarctic: The Farthest South Expedition 1907–1909* (Ware: Wordsworth Editions, 2007).

10. Hundslappadrífa, Icelandic
P. C. Headley, *The Island of Fire: Or, A Thousand Years of the Old Northmen's Home, 874–1874* (Boston: Lee & Shepard, 1875).

11. Sheen, Kashmiri
Kashmiri is recognised as a regional language in the state of Jammu and Kashmir; it is also among the twenty-two scheduled languages of India.

13. Penitentes, Spanish
Charles Darwin, *Journal of researches into the geology and natural history of the various countries visited by H.M.S. Beagle, under the command of Captain Fitz Roy, R.N., 1832 to 1836.* (London: Henry Colburn, 1839).

14. Cīruḷputenis, Latvian

George Meredith, 'The Lark Ascending', in *Poems and Lyrics of the Joy of Earth* (London: Macmillan, 1883).

15. Unatsi, Cherokee

Cherokee (known in Cherokee itself as Tsalagi or **ᏣᎳᎩ**) is an Iroquoian language, which is related to other languages like Mohawk and Seneca. In 2019 the Tri-Council of Cherokee tribes declared a state of emergency for the language, which is facing extinction. The sparrow folktale is told in *The Sparrow and the Trees: A Cherokee Folktale* (Mount Pleasant, SC: Arbordale, 2015).

18. Tykky, Finnish

J. R. R. Tolkien, letter no. 163 to W. H. Auden, 7 June 1953, in *Letters of J. R. R. Tolkien* (London: George Allen & Unwin, 1981), p. 214.

19. Barfānī chītā, Urdu

The snow leopard's 'vulnerable' status is according to the International Union for Conservation of Nature Red List of Threatened Species.

20. Snemand, Danish

'The Snow Man' is taken from Hans Christian Andersen's *Fairy Tales*, trans. Tiina Nunnally, ed. Jackie Wullschlager (New York: Viking, 2005).

'Maybe it's wrong . . .' Peter Høeg, *Miss Smilla's Feeling for Snow*, trans. Tiina Nunnally [F. David], (London: Harvill, 1993).

21. Mávro chióni, Greek

'Snow and hoar-frost . . .' Aristotle, *Meteorology*, trans. E. W. Webster, part 11, http://classics.mit.edu/Aristotle/meteorology.1.i.html.

'Anaxagoras set the appearance that snow is white . . .' *Anaxagoras of Clazomenae: Fragments and Testimonia, a Text and Translation*, trans. Patricia Curd (Toronto: University of Toronto, 2007).

'worms are found . . .' Aristotle, *History of Animals*, trans. D'Arcy Wentworth Thompson, http://classics.mit.edu/Aristotle/history_anim.html.

26. Calóg shneachta, Irish

James Joyce, *Dubliners* (London: Penguin Modern Classics, 2000).

James Joyce, *A Portrait of the Artist as a Young Man*, (London: Penguin Modern Classics, 2000).

27. Huka-rere, Māori

'fierce squalls . . .' G. Grey, *Polynesian Mythology*, illustrated edition (Christchurch: Whitcombe and Tombs, 1956).

31. Omuzira, Luganda

Roberto Montovani worked directly from Abruzzi's notes. Part of his account is available online: http://www.rwenzoriabruzzi.com/the-1906-scientific-climbing.

33. Sniegas, Lithuanian

'once these poor wretches . . .' Armand-Augustin-Louis Caulaincourt, *With Napoleon in Russia*, trans. Jean Hanoteau (New York: Dover, 2005).

36. Himá, Sanskrit

'Just when I discovered . . .' R. C. Wilson, 'Kailash Parbat and Two
   Passes of the Kumaon Himalayas', *Alpine Journal* 40, no. 230 (1928):
   23–37.

'If we conquer . . .' quoted in Peter Ellingsen, 'Scaling a Mountain
   to Destroy the Holy Soul of Tibetans', *World Tibet Network News*,
   2 June 2001.

37. Qasa, Quechua

'I believe . . .' and 'large mountains . . .' in *The Discovery and Conquest of
   Peru: Chronicles of the New World Encounter*, ed. and trans. Alexandra
   Parma Cook and Noble David Cook (Durham, NC: Duke
   University Press, 1998).

'an exceptional . . .' see https://whc.unesco.org/en/list/1459.

38. Barado, Amharic

'In one ancient inscription . . .' The text is known as the *Monumentum
   Adulitanum*.

'violence . . .' Henry Salt, *A Voyage to Abyssinia and Travels into the
   Interior of that Country* (London: Frank Cass, 1967).

39. Ttutqiksribvik, Inupiaq

'One writer recalls . . .' Susan W. Fair, 'The Northern Umiak: Shelter,
   Boundary, Identity', *Perspectives in Vernacular Architecture* 10 (2005):
   233–48.

Mizagluk, and other definitions are taken from 'Alphabetical List of
Kingikmiut Sea Ice' by Winton Weyapuk Jr, Herbert Anungazuk,
Pete Sereadlook and Faye Ongtowasruk, in *Kiŋikmi Sigum Qanuq
Ilitaavut / Wales Inupiaq Sea Ice Dictionary*, pp. 15–23. See
https://jukebox.uaf.edu/site7/sites/default/files/documents/
Preserving-our-Knowledge--Wales-Dictionary.pdf.

40. Ais i pundaun olsem kapok, Tok Pisin

Ronald Skeldon, 'Volcanic Ash, Hailstorms and Crops: Oral History
from the Eastern Highlands of Papua New Guinea', *Journal of the
Polynesian Society* 86, no. 3 (1977): 403–9.

42. Eira, Welsh

'The Welsh language is a vast drawback to Wales . . .' *Reports of the
Commissioners of Inquiry into the State of Education in Wales* (UK
Government Paper, 1846), part 2, no. 9, p. 66.

'the panoramic view of the Andes . . .' BMS 7669, quoted in Glyn
Williams, 'Welsh Contributions to Exploration in Patagonia',
Geographical Journal 135, no. 2 (1969): 213–27.

44. Pana, Inuktitut

Mitiarjuk Nappaaluk, *Sanaaq: An Inuit Novel*, trans. Bernard Saladin
D'Anglure (Winnipeg: University of Manitoba Press, 2014).

47. Hau kea, Hawaiian

For an example of a protest song, watch 'Kū ha'aheo', sung by a group of activists including footage of the *kia'i*, filmed by Oiwi TV, at https://youtu.be/F48O1qMi4ww.